Lecture Notes in Computer Science 15239

Founding Editors

Gerhard Goos
Juris Hartmanis

The series Lecture Notes in Computer Science (LNCS), including its subseries Lecture Notes in Artificial Intelligence (LNAI) and Lecture Notes in Bioinformatics (LNBI), has established itself as a medium for the publication of new developments in computer science and information technology research, teaching, and education.

LNCS enjoys close cooperation with the computer science R & D community, the series counts many renowned academics among its volume editors and paper authors, and collaborates with prestigious societies. Its mission is to serve this international community by providing an invaluable service, mainly focused on the publication of conference and workshop proceedings and postproceedings. LNCS commenced publication in 1973.

Chao Chen · Yash Singh · Xiaoling Hu
Editors

Topology- and Graph-Informed Imaging Informatics

First International Workshop, TGI3 2024
Held in Conjunction with MICCAI 2024
Marrakesh, Morocco, October 10, 2024
Proceedings

Editors
Chao Chen 🆔
Stony Brook University
Stony Brook, NY, USA

Yash Singh
Mayo Clinic
Rochester, MN, USA

Xiaoling Hu
Harvard University
Boston, MA, USA

ISSN 0302-9743   ISSN 1611-3349 (electronic)
Lecture Notes in Computer Science
ISBN 978-3-031-73966-8   ISBN 978-3-031-73967-5 (eBook)
https://doi.org/10.1007/978-3-031-73967-5

Preface

This volume presents the proceedings of the First Workshop on Topology- and Graph-Informed Imaging Informatics (TGI3), held on October 10, 2024, in Marrakesh, Morocco, in conjunction with the 27th International Conference on Medical Image Computing and Computer Assisted Intervention (MICCAI 2024).

Significant advances in computational and data science over the past decade have had an immense impact on biomedical science and healthcare. Concurrently, researchers in the biomedical fields face new challenges, mainly due to the nature of complex, often high-dimensional, noisy, and diverse datasets. All these properties of topological-based methods strongly motivate the adoption of TDA tools for various applications and domains including neuroscience, bioscience, biomedicine, and medical imaging.

This workshop focused on using TDA techniques to enhance the performance, generalizability, expressiveness, efficiency, and explainability of current methods applied to medical data. In particular, the workshop focused on using TDA tools solely or combined with other computational techniques (e.g., feature engineering and deep learning) to analyze medical data including images/videos, sounds, physiological, texts, and sequence data. The combination of TDA and other computational approaches is more effective in summarizing, analyzing, quantifying, and visualizing complex medical data.

Our workshop attracted 13 submissions worldwide, all of them undergoing a rigorous double-blind review process. The program featured four oral presentations and nine poster presentations. We were also honored to have distinguished keynote speakers at TGI3 2024 who are leaders in their respective fields, including Moo K. Chung from the University of Wisconsin–Madison.

We owe the success of TGI3 2024 to our organization committee and program committee members, whose dedication ensured the high quality of our scientific program. Their time and expertise, along with the valuable contributions of all authors, made this workshop possible.

We hope the proceedings of TGI3 2024 will serve as a valuable resource for the medical image computing community, motivating further research and innovation in addressing the challenges of exploring medical data with complex structures, topology, and geometry. We look forward to having the opportunity to continue these discussions in future workshops.

August 2024

Chao Chen
Yash Singh
Xiaoling Hu

Organization

Program Committee Chairs

Chao Chen — Stony Brook University, USA
Xiaoling Hu — Harvard Medical School, USA
Yash Singh — Mayo Clinic, USA

Editorial Chairs

Wentao Huang — Stony Brook University, USA
Meilong Xu — Stony Brook University, USA

Organization Committee

Chao Chen — Stony Brook University, USA
Johannes C. Paetzold — Imperial College London, UK
Colleen Farrelly — Post Urban Ventures, USA
Saumya Gupta — Stony Brook University, USA
Quincy Hathaway — West Virginia University, USA
Xiaoling Hu — Harvard Medical School, USA
Bjoern Menze — University of Zurich, Switzerland
Rahul Paul — Harvard Medical School, USA
Paul Rosen — University of Utah, USA
Jennifer Rozenblit — University of Texas, Austin, USA
Yash Singh — Mayo Clinic, USA

Program Committee

Colleen Farrelly — Post Urban Ventures, USA
Diana Vera Garcia — Mayo Clinic, USA
Saumya Gupta — Stony Brook University, USA
Wentao Huang — Stony Brook University, USA
Chen Li — Stony Brook University, USA
Zhenghong Li — Stony Brook University, USA
Weimin Lyu — Stony Brook University, USA

Pawan Pandey   IIT Roorkee, India
Jennifer Rozenblit   University of Texas, Austin, USA
Fan Wang   Stony Brook University, USA
Meilong Xu   Stony Brook University, USA

Contents

Deep Learning-Based Liver Vessel Separation with Plug-and-Play Modules: Skeleton Tracking and Graph Attention

Chenhao Pei, Wei Wang, Huan Zhang, Siyuan Yin, Wen Tang, Ming Meng, Weinan Xiao, and Hong Shen[✉]

Infervision Medical Technology Co., Ltd., Beijing, China
shong@infervision.com

Abstract. Accurate segmentation of liver vessels is crucial for medical applications due to its pivotal role in diagnosing liver diseases, planning surgical interventions, and assessing treatment effectiveness. In this paper, we present a new dataset for liver vessel separation and propose two novel plug-and-play modules integrated into deep learning frameworks for liver vessel segmentation. The first module, termed as the skeleton tracking module, addresses the issue of segmentation fragmentation by effectively tracking the vessel skeletons. The second module, the graph attention module, is introduced for vessel separation. We demonstrate the effectiveness of our proposed approach through comprehensive experiments, showcasing significant improvements in segmentation accuracy. The dataset is publicly available, fostering research and development. https://github.com/oneway-phil/SKTS-GAT/tree/main.

Keywords: Segmentation · Liver vessel · Skeleton Tracking · Graph

1 Introduction

Liver cancer is indeed a dangerous disease characterized by high morbidity, recurrence rates, and mortality. Its pathogenesis is complex and challenging to diagnose and treat [1]. Blood vessel segmentation is usually an important step in the diagnosis and treatment of liver cancer. Accurate vessel segmentation facilitates precise localization of pathological areas, aiding in the rational planning of surgical pathways and reducing the occurrence of intraoperative bleeding and other complications [2]. Enhanced computed tomography (CT) is one of the commonly used in liver blood vessels and has high density and resolution and rich image information. However, variations in blood flow velocity, data sampling intervals, and vascular stenosis lead to distribution changes, resulting in inconsistent grayscale depiction within the vessel area, making blood vessel segmentation difficult [14].

Recently, the field of deep learning has witnessed remarkable advancements, revolutionizing various domains, including medical imaging. It has been

---

First authors (C. Pei and W. Wang are with the same degree of contribution, they are the co-first authors).

C. Chen et al. (Eds.): TGI3 2024, LNCS 15239, pp. 1–10, 2025.
https://doi.org/10.1007/978-3-031-73967-5_1

extensively applied and has demonstrated promising results in image segmentation. However, the application of deep learning in medical image analysis faces unique challenges due to the inherent complexities of medical images. In blood vessel segmentation, vessels exhibit fine, irregular structures, presenting considerable difficulty for automated algorithms. Yuan et al. [17] propose an adaptive feature fusion network to accurately segment liver vessels from CT images, achieving higher accuracy by addressing challenges in extracting small vessels and edge vessels; Zeng et al. [18] propose a new automatic method using 3D region growing with bi-Gaussian filter and a hybrid active contour model combined with K-means clustering. By analyzing the topology of blood vessels in images to help identify and connect the branches and intersections of blood vessels, it can help improve the accuracy of segmentation algorithms. Rahman et al. [10] and Xu et al. [16] introduce novel methods for 2D medical image segmentation, using graph convolution-based decoders and feature aggregation modules.

To address the challenges in blood vessel segmentation, it is crucial to focus on the separation of different types of vessels. The accurate differentiation of arteries, veins, and capillaries is vital for various medical diagnoses and treatments, as these vessel types have distinct physiological and pathological roles. Several studies have explored different methodologies for vessel separation. Zeng et al. [19] present a novel liver-vessel segmentation and identification method that effectively distinguishes hepatic and portal veins using a distance voting mechanism. Nardelli et al. [9] introduce a novel, fully automatic approach to classifying vessels in chest CT images into arteries and veins. Xie et al. [15] propose a point-based approach for modeling complex 3D tree-shaped structures within the pulmonary system, such as airways, arteries, and veins, using high-resolution image stacks. This method preserves the graph connectivity of the tree skeleton and incorporates an implicit surface representation.

In this work, we present a new dataset for liver vessel separation and propose a novel two-stage framework with two plug-and-play modules, skeleton tracking and graph attention. The key idea is to obtain accurate liver vessel skeleton followed by vessel separation based on this skeleton. For the first stage, we propose offset and orientation losses and incorporate them to facilitate precise vessel tracking. Then, inspired by graph attention mechanisms, we introduce a novel approach, which utilizes an adjacency matrix to integrate graph attention into vessel separation. Both modules are designed to be seamlessly integrated into existing base methods, enhancing the performance of tubular structure segmentation. Our main contributions are as follows:

(1) We present a new dataset for liver vessel separation which have assembled and annotated from public datasets with a total of 146 cases.
(2) We propose a novel two-stage framework with two plug-and-play modules, Skeleton Tracking and Graph Attention.
(3) We propose offset and orientation losses to facilitate precise vessel tracking.
(4) We design a novel graph attention module, which utilizes an adjacency matrix to integrate graph attention into vessel separation.

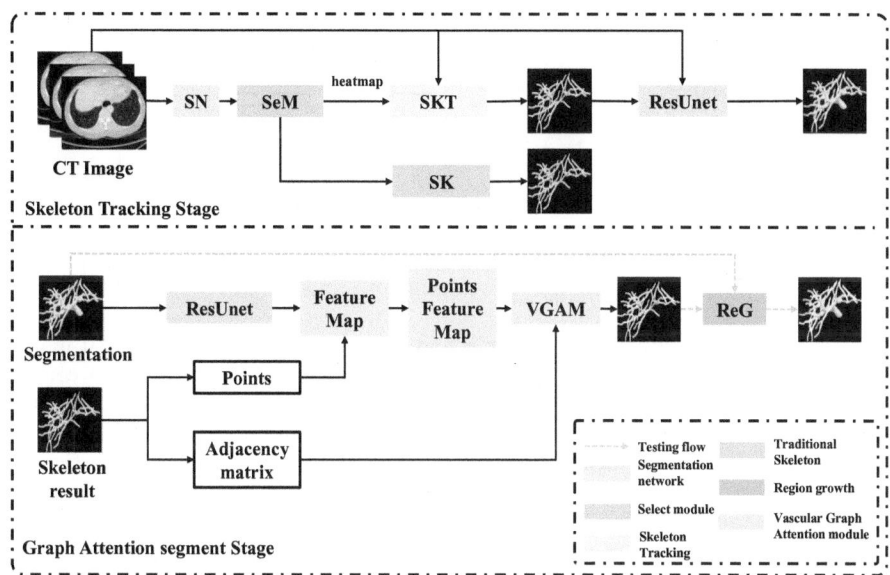

Fig. 1. Overview of the proposed two-stage framework for liver vessel segmentation. The framework comprises a skeleton tracking stage followed by a graph attention segmentation stage. More specifically, in the first stage, we process the CT image to obtain accurate vessel skeleton results, distinguishing between foreground and background. Subsequently, the graph attention segmentation stage focuses on generating segmentation outputs for various classes.

2   Method

We first formulate the problem of vessel segmentation mathematically. Formally, we have access to labeled dataset contained images and segmentation, denoted by $(X, Y) = \{(x_j, y_j)\}_{j=1}^{N}$, where N is the number of data points. We further derive the skeletal points from segmentation results and calculate the adjacency matrix to analyze the connectivity patterns within the skeletal structure, denoted by $(X, Y^S) = \{(x_j, y_j^s)\}_{j=1}^{N}$ and M. In this work, we utilize the point-graph relationships of vessel skeleton points to bridge discontinuities, enabling vessel tracking effectively.

The framework of the proposed method is illustrated in Fig. 1, containing two key processing stages. To alleviate instances of vessel segmentation discontinuity, we introduce a vessel tracking module in the first stage, as illustrated in Sect. 2.1. Further in Sect. 2.2, we introduce a graph attention network to achieve vessel separation.

2.1   Skeleton Tracking Stage(SKTS)

The skeleton tracking stage is designed to obtain segmentation results for a vessel's two-class anatomical structure from CT images is illustrated in Fig. 1.

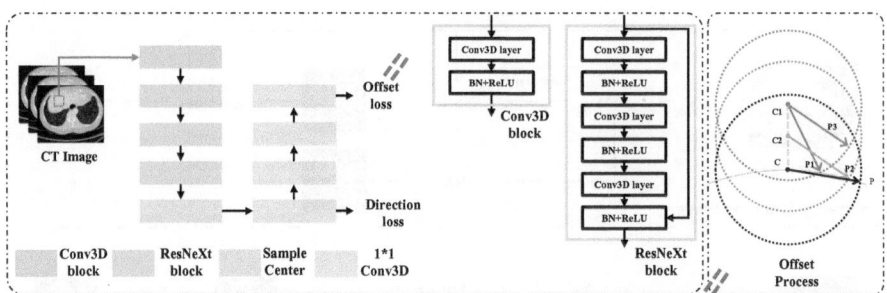

Fig. 2. The left is the architecture of SKN and the right is the offset process. It mainly includes 1 conv3D block, 3 ResNeXt blocks, 1 Sample Center block and 4 1×1 conv3D blocks. Point C on the centerline, point P is the target point, CP is the direction to point C, C_1 is the offset point, and C_1P_1 is the offset direction. Point C_2 is the point that meets the conditions among points C and C_1, and C_2P_2 is the final direction.

Giving a CT image x, the SKTS first generates a probability map of vessel presence using a base arbitrary segmentation network. If the proportion of vessel pixels with a confidence probability exceeding 0.5 divided by exceeding 0.01 surpasses 85%, the traditional skeleton extraction method is employed directly. Otherwise, vessel tracking is performed. The tracked vessels exhibit more complete structures after tracing. Subsequently, the tracked vessels, along with the original CT image, are fed into ResUnet [20] for segmentation, resulting in more accurate vessel segmentation outcomes.

For SN(segmentation network) and ResUnet, we use either Unet or nnUNet as the SN. Both SN and ResUnet are independently trained to accomplish the 2-class segmentation task for foreground and background. The training processes follow the original setting.

For the vessel tracking(SKT) process, we first select candidate segmentation maps from the heatmap, where the confidence probability exceeds h. These candidate segmentation maps undergo traditional skeleton extraction algorithms to obtain a candidate seed set. Points within this candidate seed set, having more than 2 neighbors within a 26-neighborhood, are selected as the starting points for tracking, with a tracking step size set to 1. And tracking concludes upon reaching the vicinity of a skeleton point or when the tracking point reaches the lower probability value of the heatmap. To obtain tracking directions for each point, we introduce a skeleton direction network(SKN). Figure 2 shows the architecture of the SKN. It mainly includes 1 conv3D block, 3 ResNeXt blocks, 1 Sample Center block and 4 1×1 conv3D blocks. Sample Center block aims to extract features from the central positions of the 3D feature map.

In practical situations, some center points may experience offsets which affect model performance. We introduce offset loss to smoothly rectify previous offsets, as shown in Fig. 2. When given a point C on the centerline, the intersection of the ball (centered on C with a radius of 1) and the centerline is denoted as point P. Ordinarily, the direction CP is predicted and connected with a radius

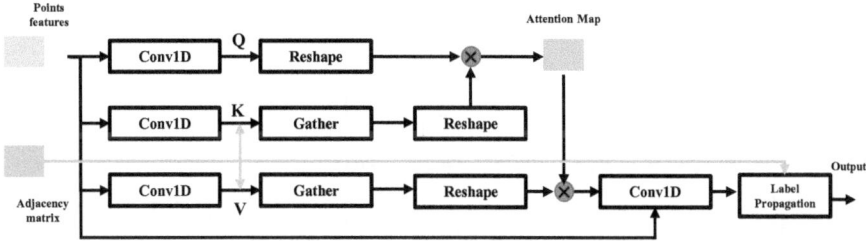

Fig. 3. The architecture of VGAM. It consists of multiple linear transformations applied to the input features to compute queries(Q), keys(K), and values(V).

to determine the final prediction. However, to mitigate the risks of overfitting and prediction bias, we employ offset augmentation during training. Specifically, through offset augmentation, we derive point C_1. Excessive offset can result in significant deviation between the predicted direction and the true value, thereby leading to an unsmooth predicted centerline. Hence, in such scenarios, it is prudent to select a moderate point C_2 direction as the label to ensure a smoother prediction process. During training, we deliberately introduce a soft deviation coefficient to adjust the center point and incorporate regression loss to train the model. The training loss of the direction $\mathcal{L}_D$ and offset $\mathcal{L}_{Off}$ can be defined as,

$$\mathcal{L}_D = -\frac{1}{N} \sum_{i=1}^{N} \sum_{j=1}^{C} y_{ij}^d \log(\hat{y}_{ij}^d), \tag{1}$$

where N is the number of samples, C is the number of classes (we use 1000, 1000 directions), y_{ij}^d is the ground truth label for sample i and class j, and $\hat{y}_{ij}^d$ is the predicted for sample i and class j.

$$\mathcal{L}_{off} = L_{smooth\_L1}(y^{off}, \hat{y}^{off}), \tag{2}$$

where L_{smooth_L1} is smooth L1 loss, y^{off} is the ground truth label and $\hat{y}^{off}$ is the predictiction. Finally, the overall loss function of the SKT is,

$$\mathcal{L}_{SKT} = \lambda_D \mathcal{L}_D + \lambda_{off} \mathcal{L}_{off}, \tag{3}$$

2.2 Graph Attention Segment Stage(GATSS)

In the Graph Attention Segment Stage(GATSS), complementary information from points and feature maps are expected to be fused and boost the segmentation performance. Specifically, we construct a branch to capture the point feature map from the image feature map. The points and adjacency matrix are sampled and computed from the skeleton results, respectively. Then the adjacency matrix and the feature map of points are fed to the vessel graph attention module(VGAM). In the training process, in order to reduce the memory usage,

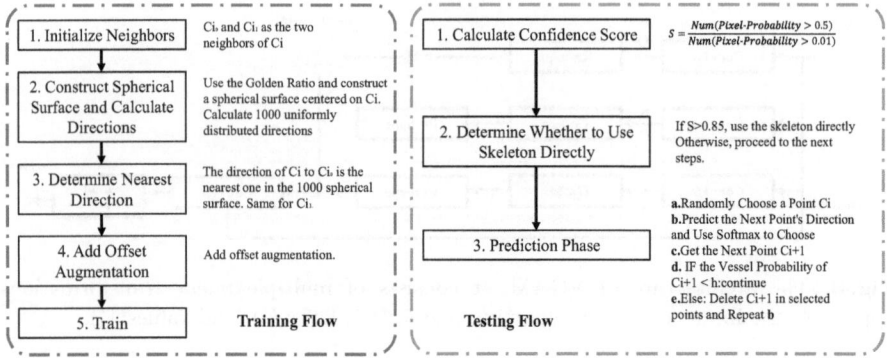

Fig. 4. The Training and testing flow.

we only train some qualified nodes in the adjacency matrix. In the prediction process, we use the model prediction results combined with the adjacency matrix for label propagation [11] to obtain the final result.

Figure 3 shows the architecture of VGAM. It consists of multiple linear transformations applied to the input features to compute queries, keys, and values. The queries (Q), keys (K), and values (V) are then reshaped to facilitate attention mechanism computations. The attention weights are calculated using a softmax function applied to the dot product of queries and keys. These attention weights are then used to weight the values and obtain the updated node representations. Finally, the updated node representations are passed through a linear layer and added to the original features before applying a ReLU activation function. This process effectively captures the relationship between nodes in the graph and produces enhanced node representations. The VGAM process can be described as the following formula,

$$VGAM(f_p, M) = \mathcal{W}_v(f_p, M) \cdot \sigma(\mathcal{W}_q(f_p) \cdot \mathcal{W}_k(f_p, M)) + f_p, \qquad (4)$$

where M is the adjacency matrix and f_p is the features of points. $\mathcal{W}_q, \mathcal{W}_k, \mathcal{W}_v$ are the parameters of the multiple linear transformations. σ is sofmax function.

To enhance the performance of the model, we introduce a segmentation branch into the feature map. The output after VGAM is the sparse point segmentation result. The associated loss is as follows,

$$\mathcal{L}_{VGAM} = \mathcal{C}(y_i, \widehat{y}_i) + \beta \cdot Dice(y_i, \widehat{y}_i) + \kappa \mathcal{C}(y_i^s, \widehat{y}_i^s), \qquad (5)$$

where $\widehat{y}_i$ is the segmentation prediction from the segmentaion of SKTS, $\mathcal{C}(y_i, \widehat{y}_i)$ is the cross-entropy loss, $Dice(y_i, \widehat{y}_i)$ is the Dice loss, and β, κ are the hyperparameter to balance them. $\widehat{y}_i^s$ is the point segmentation prediction from the segmentation of SKTS.

Table 1. The Dice and Dice$_{CL}$ values of ablation study for the proposed p^{2}SG. $^\diamond$ means that using UNet as SN. SKTS: Skeleton Tracking Stage; VGAM: vessel graph attention module; Without VGAM, we use GCN to replace it.

Method	SKTS	VGAM	IPV		IV		Mean	
			Dice	Dice$_{CL}$	dice	Dice$_{CL}$	dice	Dice$_{CL}$
p^{2}SG$^\diamond$ w/o SKTS	×	✓	0.7759 ± 0.0018	0.8776 ± 0.0023	0.7600 ± 0.0039	0.8853 ± 0.0023	0.7660 ± 0.0029	0.8785 ± 0.0024
p^{2}SG$^\diamond$ w/o VGAM	✓	×	0.7979 ± 0.0013	0.8717 ± 0.0016	0.7709 ± 0.0036	0.8783 ± 0.0025	0.7844 ± 0.0027	0.8750 ± 0.0020
p^{2}SG$^\diamond$	✓	✓	0.8115 ± 0.0010	0.9040 ± 0.0011	0.7868 ± 0.0022	0.9066 ± 0.0011	0.7991 ± 0.0018	0.9053 ± 0.0011

Table 2. The Dice and Dice$_{CL}$ values for liver vessel segmentation with CT images. $^\diamond$ means that using UNet as SN. * means that using nnUnet as SN.

Method	IPV		IV		Mean	
	Dice	Dice$_{CL}$	Dice	Dice$_{CL}$	Dice	Dice$_{CL}$
UNet [6]	0.8020 ± 0.0014	0.8949 ± 0.0010	0.7787 ± 0.0033	0.8990 ± 0.0033	0.7903 ± 0.0025	0.8969 ± 0.0022
DenseUNet [3]	0.7729 ± 0.0036	0.8496 ± 0.0056	0.7566 ± 0.0075	0.8654 ± 0.0086	0.7647 ± 0.0057	0.8575 ± 0.0073
WingsNet [21]	0.7183 ± 0.0040	0.6772 ± 0.0014	0.6500 ± 0.0145	0.7653 ± 0.0200	0.6842 ± 0.0106	0.7212 ± 0.0051
AvdNet [13]	0.7544 ± 0.0019	0.8219 ± 0.0030	0.7333 ± 0.0066	0.8358 ± 0.0070	0.7439 ± 0.0044	0.8288 ± 0.0051
nnUNet [7]	0.8415 ± 0.0011	0.9227 ± 0.0010	0.8257 ± 0.0029	0.9192 ± 0.0026	0.8335 ± 0.0021	0.9209 ± 0.0018
p^{2}SG$^\diamond$	0.8115 ± 0.0010	0.9040 ± 0.0011	0.7868 ± 0.0022	0.9066 ± 0.0011	0.7991 ± 0.0018	0.9053 ± 0.0011
p^{2}SG*	0.8419 ± 0.0012	0.9210 ± 0.0010	0.8333 ± 0.0021	0.9253 ± 0.0012	0.8376 ± 0.0016	0.9232 ± 0.0011

3   Experiment

3.1   Dataset

We evaluate our method on liver vessel segmentation tasks using the dataset we assembled and annotated. The dataset was collected from publicly available sources, including 3D-IRCADb [12], CHAOS Challenge [8], LiVS [4], MSD(Task08 HepaticVessel) [1] and Sliver07 [5], a total of 146 cases were included in the dataset with portal venous phase, layer thickness less than 2 mm and vascular imaging greater than level 3. These data were manually annotated to provide segmentation labels for intrahepatic veins(IV), and intrahepatic portal veins(IPV). Veins were annotated to level 3. Portal veins were annotated to level 4. The dataset was then randomly split into training and testing sets with a ratio of 4:1, resulting in 117 cases for training and 29 cases for testing.

3.2   Implementations

We trained our models with 3D CT images. We choose UNet or nnUNet as our SN for the binary classification task(foreground and background) and follow the original settings. The ResUnet in SKTS also follows the original settings. The models of SKT and GATSS were implemented in Python and optimized by using the Adam algorithm.

For the SKT in SKTS: In each training iteration, the segmentation form SN was fed into the SKT. For data augmentation, we apply random offsets and rotations. The SKT can be trained by minimizing the direction loss $\mathcal{L}_D$ and offset loss $\mathcal{L}_{off}$.

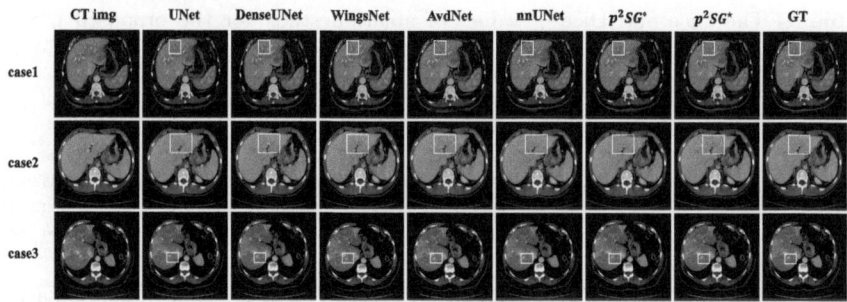

Fig. 5. Visualization of segmentation results of three CT cases. The structures of IPV and IV are in red and green colors, respectively. Blue boxes show regions of interest. $^{\diamond}$ means that using UNet as SN. * means that using nnUnet as SN. (Color figure online)

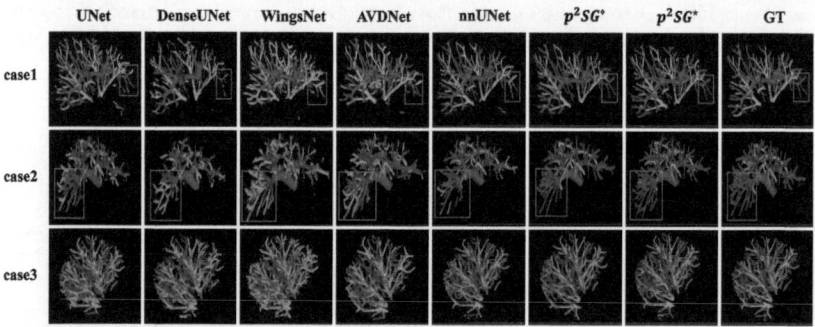

Fig. 6. Visualization of 3D segmentation results of three CT cases. The structures of IPV and IV are in red and green colors, respectively. Blue boxes show regions of interest. $^{\diamond}$ means that using UNet as SN. * means that using nnUnet as SN. (Color figure online)

For GATSS: The hyper-parameter β and κ were set to 1 and 1, respectively. Finally, the network can be trained by minimizing the $\mathcal{L}_{VGAM}$.

We used the Dice coefficient (Dice) and the CL Dice coefficient (Dice$_{CL}$) to evaluate the segmentation accuracies.

Figure 4 provide the training and testing flow.

3.3 Ablation Study

The proposed p^{2}SG utilizes three major techniques to improve the performance, including SKTS and VGAM. To validate their effectiveness, we performed three ablation studies. Table 1 summarizes the results of these studies, where '×' and '√' respectively indicate the exclusion and inclusion of the module.

The skeleton tracking stage is designed to obtain segmentation results for a vessel's two-class anatomical structure from CT images, we implemented the variant of p^{2}SG without this technique. As Table 1 shows, with the SKTS, the performance of p^{2}SG obtained better results with the mean of average Dice score being 0.7991 versus 0.7660 (p < 0.01)and average CL Dice score being 0.9053 versus 0.8785 (p < 0.01). Due to the more complex vessel structure of the IPV, the p^{2}SG achieved a greater improvement in IPV than IV. The SKTS effectively achieves more comprehensive and reliable vessel segmentation, especially when dealing with the intricate vascular network of the IPV.

GCNs can effectively be applied to vessel segmentation tasks that capture spatial and structural relationships. Attention can improve the integration of local and global contextual information, enhancing the segmentation accuracy. As Table 1 shows, compared to GCN, the p^{2}SG with VGAM improved the average Dice score and Dice_{CL} score by about 0.0107 and 0.0303.

3.4 Comparison Study

The results are presented in Table 2. p^{2}SG achieved the best performance in both Dice and Dice_{CL} values. With the design for tubular structures, p^{2}SG performed better in Dice, with a margin about 0.0937 (p < 0.001)than Avd-Net [13] and 0.1534 (p < 0.001)than WingsNet [21]. Compared to nnUNet [7], p^{2}SG* improved the Dice and Dice_{CL} of the liver vessel segmentation with about 0.0041 (p = 0.7063) and 0.0023 (p = 0.8102). The proposed p^{2}SG* shows only a slight improvement compared to nnUNet. This is because nnUNet already delivers excellent vessel segmentation results with few instances of vessel breakage. As a result, the effectiveness of the SK is significantly reduced, given that the primary advantage of SK lies in its ability to address and correct segmentation discontinuities, which are already minimal in nnUnet's outputs.

Figure 5 and 6 provide three examples of liver vessel segmentation for illustration about 2D and 3D, respectively. The visualized results of the proposed method demonstrate a more complete and continuous vessel segmentation, highlighting its effectiveness in capturing the intricate vessel structures within the liver.

4 Conclusion

In this work, we have presented a new dataset for liver vessel separation and proposed a novel framework for liver vessel segmentation. Different from the previous method, our method, referred to as p^{2}SG, can effectively track the vessel skeleton and use the graph structure characteristics of blood vessels. Compared with the state-of-the-art segmentation methods, the proposed p^{2}SG demonstrated superior performance and promising results.

References

1. Antonelli, M., et al.: The medical segmentation decathlon. Nat. Commun. **13**(1), 4128 (2022)
2. Aoki, T., Kubota, K.: Preoperative portal vein embolization for hepatocellular carcinoma: consensus and controversy. World J. Hepatol. **8**(9), 439 (2016)

3. Cai, S., Tian, Y., Lui, H., Zeng, H., Wu, Y., Chen, G.: Dense-UNet: a novel multi-photon in vivo cellular image segmentation model based on a convolutional neural network. Quant. Imaging Med. Surg. **10**(6), 1275 (2020)
4. Gupta, A., Dollar, P., Girshick, R.: LVIS: a dataset for large vocabulary instance segmentation. In: Proceedings of the IEEE/CVF Conference on Computer Vision and Pattern Recognition, pp. 5356–5364 (2019)
5. Heimann, T., et al.: Comparison and evaluation of methods for liver segmentation from CT datasets. IEEE Trans. Med. Imaging **28**(8), 1251–1265 (2009)
6. Huang, H., et al.: UNet 3+: A full-scale connected UNet for medical image segmentation. In: ICASSP 2020-2020 IEEE International Conference on Acoustics, Speech and Signal Processing (ICASSP), pp. 1055–1059. IEEE (2020)
7. Isensee, F., Jaeger, P.F., Kohl, S.A., Petersen, J., Maier-Hein, K.H.: nnU-Net: a self-configuring method for deep learning-based biomedical image segmentation. Nat. Methods **18**(2), 203–211 (2021)
8. Kavur, A.E., et al.: Chaos challenge-combined (CT-MR) healthy abdominal organ segmentation. Med. Image Anal. **69**, 101950 (2021)
9. Nardelli, P., et al.: Pulmonary artery-vein classification in CT images using deep learning. IEEE Trans. Med. Imaging **37**(11), 2428–2440 (2018)
10. Rahman, M.M., Marculescu, R.: G-CASCADE: efficient cascaded graph convolutional decoding for 2D medical image segmentation. In: Proceedings of the IEEE/CVF Winter Conference on Applications of Computer Vision, pp. 7728–7737 (2024)
11. Selle, D., Preim, B., Schenk, A., Peitgen, H.O.: Analysis of vasculature for liver surgical planning. IEEE Trans. Med. Imaging **21**(11), 1344–1357 (2002)
12. Soler, L., et al.: 3D image reconstruction for comparison of algorithm database **3** (2010). https://www.ircad.fr/research/datasets/liver-segmentation-3d-ircadb-01
13. Wang, W., et al.: AVDNet: joint coronary artery and vein segmentation with topological consistency. Med. Image Anal. **91**, 102999 (2024)
14. Wetzel, S.G., Kirsch, E., Stock, K.W., Kolbe, M., Kaim, A., Radue, E.W.: Cerebral veins: comparative study of CT venography with intraarterial digital subtraction angiography. Am. J. Neuroradiol. **20**(2), 249–255 (1999)
15. Xie, K., Yang, J., Wei, D., Weng, Z., Fua, P.: Efficient anatomical labeling of pulmonary tree structures via implicit point-graph networks. arXiv preprint arXiv:2309.17329 (2023)
16. Xu, S., Duan, L., Zhang, Y., Zhang, Z., Sun, T., Tian, L.: Graph-and transformer-guided boundary aware network for medical image segmentation. Comput. Methods Programs Biomed. **242**, 107849 (2023)
17. Yuan, Y., et al.: AFF-NET: an adaptive feature fusion network for liver vessel segmentation from CT images. In: 2023 IEEE 20th International Symposium on Biomedical Imaging (ISBI), pp. 1–5. IEEE (2023)
18. Zeng, Y.z., et al.: Automatic liver vessel segmentation using 3d region growing and hybrid active contour model. Comput. Biol. Med. **97**, 63–73 (2018)
19. Zeng, Y.z., et al.: Liver vessel segmentation and identification based on oriented flux symmetry and graph cuts. Comput. Methods Programs Biomed. **150**, 31–39 (2017)
20. Zhang, Z., Liu, Q., Wang, Y.: Road extraction by deep residual U-Net. IEEE Geosci. Remote Sens. Lett. **15**(5), 749–753 (2018)
21. Zheng, H., et al.: Alleviating class-wise gradient imbalance for pulmonary airway segmentation. IEEE Trans. Med. Imaging **40**(9), 2452–2462 (2021)

ccDice: A Topology-Aware Dice Score Based on Connected Components

Pierre Rougé[1,2]($\boxtimes$) (ID), Odyssée Merveille[2] (ID), and Nicolas Passat[1] (ID)

[1] Université de Reims Champagne Ardenne, CRESTIC, Reims, France
`pierre.rouge@creatis.insa-lyon.fr`
[2] Univ Lyon, INSA-Lyon, Université Claude Bernard Lyon 1, UJM-Saint Etienne, CNRS, Inserm, CREATIS UMR 5220, U1294, 69100 Lyon, France

Abstract. Image segmentation is a complex task that aims to simultaneously satisfy various quality criteria. In this context, topology is being increasingly considered. Guaranteeing correct topological properties is indeed crucial for objects presenting challenging shapes. Designing topology-aware metrics is then relevant, both for assessing the quality of segmentation results and for designing losses involved in learning procedures. In this article, we introduce ccDice (connected component Dice), a topological metric that generalises the popular Dice score. By contrast to Dice, that acts at the scale of pixels, ccDice acts at the scale of connected components of the compared objects, leading to a topological assessment of their relative structure and embedding. ccDice is a simple, explainable, normalized and low-computational topological metric. We provide a formal definition of ccDice, an algorithmic scheme for computing it, and we assess its behaviour by comparison to other usual topological metrics. Code is available on GitHub: https://github.com/PierreRouge/ccDice.

Keywords: Topology · Metric · Binary objects · Connectedness · Segmentation

1 Introduction

Segmentation is one of the most crucial tasks in medical image analysis. The result of a segmentation process is a set of pixels that represents a structure of interest (e.g. organ, tissue, lesion). This digital object should exhibit correct properties with respect to the real structure it describes, in terms of morphology, geometry and topology.

Early in the raise of computer science, various approaches were proposed to design topological models for numerical imaging [12,18] with the purpose to be compliant with the underlying continuous topology of Euclidean spaces [13,14]. This opened the way to the development of a rich panel of tools for

Supplementary Information The online version contains supplementary material available at https://doi.org/10.1007/978-3-031-73967-5_2.

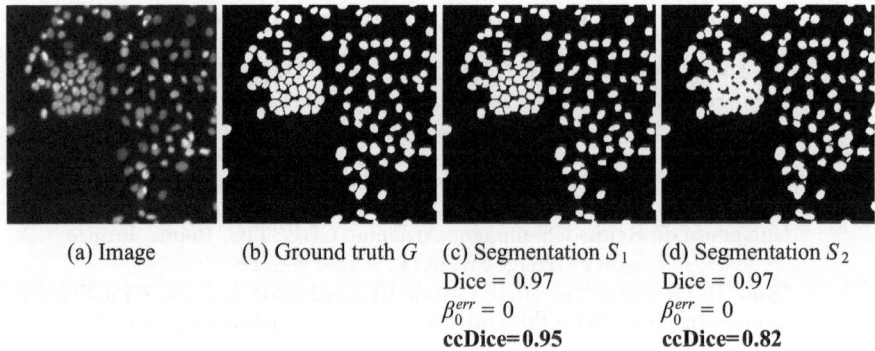

(a) Image   (b) Ground truth G (c) Segmentation S_1 (d) Segmentation S_2

Dice = 0.97   Dice = 0.97

$\beta_0^{err} = 0$ $\beta_0^{err} = 0$

ccDice=0.95   **ccDice=0.82**

Fig. 1. An image of cells (a), its ground truth G (b) and two segmentation results S_1 (c) and S_2 (d). The cluster of cells is incorrectly connected in S_2, which also contains false positive artifacts, resulting in an identical number of CCs as in (b–c), and thus a null β_0^{err} metric. Although the false positives and false negatives are not at the same locations, their total number is equal in both segmentations, resulting in equal Dice scores (0.97). Based solely on Dice and β_0^{err}, it is impossible to differentiate these segmentations, even though S_1 is better than S_2. By contrast, ccDice rates S_1 higher than S_2 (0.95 vs 0.82).

digital objects, based on topological invariants (e.g. connectedness [17], homology [7], homotopy [5]). These topological concepts were progressively involved in the design of methods and tools dedicated to medical image analysis and processing [19]. For decades, topological notions were directly embedded in algorithms to guide them by modeling topological priors or providing topological regularization schemes. The rapid development of deep learning in medical image segmentation, together with the growing trend to consider topology as a quality feature, led to a new paradigm. Topological concepts now tend to be embedded in metrics, for designing both quality scores dedicated to the evaluation of segmentation results, and losses dedicated to train topology-aware machine/deep learning models.

Therefore, recent works have introduced novel metrics specifically designed to take into account topological information. In the context of tubular structure segmentation, Shit et al. [21] proposed the centerline Dice (clDice), which considers the medial axes of both the predicted and ground truth segmentation to avoid bias effects induced by larger vs smaller structures. Some important works leverage discrete Morse theory (DMT) to identify important topological structures from the likelihood map outputted by a neural network. For example, Hu et al. [11] designed a loss focusing on the correct segmentation of these critical structures. Also, Gupta et al. [9] introduced a probabilistic DMT framework which allows computing a structure-wise uncertainty estimation which is more interpretable than pixel-wise uncertainty, especially for curvilinear structures. Algebraic topology, including topological descriptors, such as Betti numbers or persistent barcodes [1], were also investigated. For instance, Clough et al. [4]

proposed to include topological priors in the segmentation. Their loss function enforces the persistent barcodes of the predicted segmentation to be in line with the theoretical Betti numbers of the target object. Byrne et al. [2] extended this approach to the case of multiclass segmentation. Alternatively, Perret et al. [16] proposed to rely on morphological trees in order to model topological features related to the first and last homology groups, leading to a differentiable, low-cost topological loss likely to constrain the structure of a grey-level image. Hu et al. [10] introduced a loss enforcing the segmentation to have the same Betti numbers as the ground truth. However, these topological descriptors lack spatial awareness. Indeed, two images with identical topological descriptors can represent significantly different structures. To address this issue, Stucki et al. [22] leveraged persistence barcodes and designed a loss function that spatially matches objects with the same topological features between the predicted and ground truth segmentations. This approach marked a significant progress as it proposed a topological and spatially coherent matching of observed features. A drawback of this approach is its high computational cost, which makes it hardly tractable in many cases, in particular in 3 dimensions.

In this article, we propose a novel metric called ccDice, for *connected component Dice*, to assess both the topological and spatial accuracy of a segmentation. We designed a metric being (1) explainable, (2) fast to compute, (3) related to both the topology and the spatial embedding of the objects. The notion of connected components (CCs), that provides an intermediate level of topological representation of an object, is a good trade-off regarding objectives (1–3). ccDice is a generalization of the usual Dice score [6] and thus shares its relevant properties: an explainable and normalized metric. The core idea is to design a mapping between the CCs of the prediction and those of the ground-truth. This mapping is driven by the spatial matching between the CCs, thus involving spatial embedding in the topological analysis. The behaviour of ccDice is illustrated in Fig. 1.

We provide a formal definition of ccDice in Sect. 2. In Sect. 3, we describe an algorithm to compute efficiently ccDice. We experimentally evaluate the relevance of ccDice in Sect. 4, by comparison with other classic overlap and topological metrics.

2   From Dice to ccDice

We aim to compare binary images based on their CCs. This requires to embed images in a topological space (e.g. [12,18]). Connected components will be handled as elements of a partition of an image. Thus ccDice is valid for digital images, but more generally for any set subdivided as a partition. Given a (nonempty) binary image X, the set of its CCs is noted $\mathcal{C}[X]$ and is a partition of X.

2.1   Dice from (Mis)matching

Let $S, G \subseteq \Omega$ be two subsets of a set Ω (the support of the image), that may represent a segmentation result (S) and ground truth (G). The Dice score [6] of S with respect to G is defined as:

$$Dice(S, G) = \frac{2\,S \cap G|}{|S| + |G|} . \tag{1}$$

By noting $tp(S, G) = |S \cap G|$, $fp(S, G) = |S \setminus G|$ and $fn(S, G) = |G \setminus S|$, Eq. (1) can be written as follows:

$$Dice(S, G) = \frac{2 \cdot tp(S, G)}{2 \cdot tp(S, G) + fp(S, G) + fn(S, G)} . \tag{2}$$

Alternatively to this definition in terms of true positives (tp), falsepositives(fp) and false negatives (fn), we can view the Dice score as a combination of the number of matching pixels (m) and mismatching pixels $(\overline{m})$ between S and G:

$$Dice(S, G) = \frac{m(S, G) + m(G, S)}{m(S, G) + \overline{m}(S, G) + m(G, S) + \overline{m}(G, S)} = \frac{m(S, G) + m(G, S)}{|S| + |G|}, \tag{3}$$

where:

$$m(S, G) = m(G, S) = |S \cap G| = tp(S, G) = tp(G, S) \tag{4}$$
$$\overline{m}(S, G) = |S \setminus G| = fp(S, G) = fn(G, S) = |S| - m(S, G). \tag{5}$$

2.2   Matching Connected Components

Our purpose is to extend the notion of Dice from point comparison to CC comparison. Following Eq. (3), this requires to generalize the notions of (mis)matching (Eqs. (4 and 5)) from sets of points to sets of CCs. Intuitively, a CC X of the segmentation S matches a CC Y of the ground truth G if their intersection is significant with respect to the size of X.

Let $\varphi_{S,G}^{\lambda} : \mathcal{C}[S] \to \mathcal{C}[G]$ be the matching function of CCs from S to a set G, where $\lambda \in (0, 1]$ is a parameter controlling the required degree of the overlap. We define the *embedding score*, $\mathcal{E}(X, Y)$, as a function quantifying the degree of overlap of X with respect to Y:

$$\mathcal{E}(X, Y) = \frac{|X \cap Y|}{|X|} \in [0, 1] . \tag{6}$$

Then, the matching $\varphi_{S,G}^{\lambda}$ is defined such that for any $X \in \mathcal{C}[S]$, we have:

$$\mathcal{E}(X, \varphi_{S,G}^{\lambda}(X)) \geqslant \lambda . \tag{7}$$

Remark 1. *A CC $X \in \mathcal{C}[S]$ has an image by $\varphi_{S,G}^{\lambda}$ in $\mathcal{C}[G]$ iff there exists a CC $Y \in \mathcal{C}[G]$ such that $\mathcal{E}(X, Y) \geqslant \lambda$.*

Remark 2. *If* $\lambda > 0.5$ *then* $\varphi_{S,G}^{\lambda}$ *is unique.*

With this definition, each CC $X \in \mathcal{C}[S]$ is associated to *at most one* CC $Y \in \mathcal{C}[G]$. However, each CC $Y \in \mathcal{C}[G]$ could be associated to *many* CCs $X \in \mathcal{C}[S]$. This property is undesirable as we need each CC $Y \in \mathcal{C}[G]$ to be associated with *at most one* CC $X \in \mathcal{C}[S]$. To enforce this behavior, we require $\varphi_{S,G}^{\lambda}$ to be injective (see Sect. 3).

We can now define the number of matching (μ) and of mismatching ($\overline{\mu}$) on CCs as follows:

$$\mu(S, G) = |\{X \in \mathcal{C}[S] \mid \exists Y \in \mathcal{C}[G], Y = \varphi_{S,G}^{\lambda}(X)\}| \tag{8}$$

$$\overline{\mu}(S, G) = |\{X \in \mathcal{C}[S] \mid \forall Y \in \mathcal{C}[G], Y \neq \varphi_{S,G}^{\lambda}(X)\}| = |S| - \mu(S, G) . \tag{9}$$

2.3   ccDice from Matching

We are now ready to define the notion of ccDice. This definition consists of substituting in Eq. (3) the number of (mis)matchings m and $\overline{m}$ on pixels (Eqs. (4 and 5)) by the number of (mis)matchings μ and $\overline{\mu}$ on CCs (Eqs. (8 and 9)). We set:

$$ccDice(S, G) = \frac{\mu(S, G) + \mu(G, S)}{\mu(S, G) + \overline{\mu}(S, G) + \mu(G, S) + \overline{\mu}(G, S)} = \frac{\mu(S, G) + \mu(G, S)}{|\mathcal{C}[S]| + |\mathcal{C}[G]|} . \tag{10}$$

Remark 3. *We have* $ccDice(S, G) \in [0, 1]$.

Remark 4. *We have* $ccDice(S, G) = 1$ *iff* $|\mathcal{C}[S]| = |\mathcal{C}[G]|$ *and both* $\varphi_{S,G}^{\lambda}$ *and* $\varphi_{G,S}^{\lambda}$ *are bijective.*

The notion of ccDice generalizes the notion of Dice.

Proposition 5. *If we endow* Ω *with a totally disconnected topological space (i.e. a space that has only singletons as connected subsets), then for two nonempty sets* $S, G \subseteq \Omega$, *we have* $Dice(S, G) = ccDice(S, G)$.

3   Computing ccDice

Let $S, G \subseteq \Omega$ be two nonempty sets. We note $\mathcal{C}[S] = \{X_i\}_{i=1}^{t}$ and $\mathcal{C}[G] = \{Y_j\}_{j=1}^{u}$ with $t, u \geqslant 1$, the sets of CCs of S and G, respectively. The computation of ccDice (Algorithm 1) with respect to S and G mainly consists of building the two injective functions $\varphi_{S,G}^{\lambda} : \mathcal{C}[S] \to \mathcal{C}[G]$ and $\varphi_{G,S}^{\lambda} : \mathcal{C}[G] \to \mathcal{C}[S]$ in order to set the values $\mu(S, G)$ and $\mu(G, S)$. These two functions depend on the parameter $\lambda \in (0, 1]$ that determines the tolerance of the embedding between the compared CCs (Eqs. (6 and 7)).

The computation of $\varphi_{S,G}^{\lambda}$ and $\varphi_{G,S}^{\lambda}$ is described in Algorithm 2. In practice, we process each candidate pair $(X_i, Y_j) \in \mathcal{C}[S] \times \mathcal{C}[G]$ for which we have precomputed the embedding score $\varepsilon_{i,j} = \mathcal{E}(X_i, Y_j)$. We add a pair (X_i, Y_j) to

---

Algorithm 1: Compute ccDice

   Input: $S, G \subseteq \Omega$; $\lambda \in (0, 1]$
   Output: ccDice $\in [0, 1]$
1 Build $\mathcal{C}[S]$; $\mathcal{C}[G]$
2 $\mu(S, G) :=$ Compute Matching$(\mathcal{C}[S], \mathcal{C}[G], \lambda)$
3 $\mu(G, S) :=$ Compute Matching$(\mathcal{C}[G], \mathcal{C}[S], \lambda)$
4 ccDice $:= (\mu(S, G) + \mu(G, S))/(|\mathcal{C}[S]| + |\mathcal{C}[G]|)$

Algorithm 2: Compute matching

   Input: $\mathcal{C}[S] = \{X_i\}_{i=1}^{t}$; $\mathcal{C}[G] = \{Y_j\}_{j=1}^{u}$; $\lambda \in (0, 1]$
   Output: $\mu(S, G)$
1 Build $\{\varepsilon_{i,j}\}_{(i,j)\in[\![1,t]\!]\times[\![1,u]\!]}$
2 Sort $E = \{(i, j) \mid \varepsilon_{i,j} \geqslant \lambda\}$ **by decreasing values of** $\varepsilon_{i,j}$
3 $\mu(S, G) := 0$
4 foreach $(i, j) \in E$ **(sorted) do**
5   **if** $(i, \star)$ **and** $(\star, j)$ **are not discarded then**
6 $\mu(S, G) := \mu(S, G) + 1$
7   **Discard** $(i, \star)$
8   **Discard** $(\star, j)$

$\varphi_{S,G}^{\lambda}$, i.e. we set $\varphi_{S,G}^{\lambda}(X_i) = Y_j$ and increment $\mu(S, G)$ if $\mathcal{E}(X_i, Y_j) \geqslant \lambda$. When doing so, the other pairs $\mathcal{E}(X_i, \star)$ can no longer be considered (since $\varphi_{S,G}^{\lambda}$ is a function) and the same holds for the pairs $\mathcal{E}(\star, Y_j)$ (since $\varphi_{S,G}^{\lambda}$ is injective). As stated in Remark 2, when $\lambda > 0.5$, the definition of $\varphi_{S,G}^{\lambda}$ and $\varphi_{G,S}^{\lambda}$ is unique. By contrast, when $\lambda \leqslant 0.5$, many valid functions may be defined. In order to deal with this indeterminism, when building $\varphi_{S,G}^{\lambda}$ we process the candidate pairs $(X_i, Y_j) \in \mathcal{C}[S] \times \mathcal{C}[G]$ by decreasing value of $\mathcal{E}(X_i, Y_j)$ (the same holds for $\varphi_{G,S}^{\lambda}$). A complexity analysis (presented in supplementary materials) shown that the overall space and time costs of Algorithm 1 are $\mathcal{O}(n)$ and $\mathcal{O}(n \log n)$, respectively.

4 Evaluation

4.1 Experiments

We compare ccDice with standard metrics dedicated to segmentation assessment: (1) the Dice score [6] which is the gold standard for pixel-wise segmentation; (2) β_0^{err}, the absolute difference between the number of CCs of the segmentation and the ground-truth; (3) clDice [21] which is dedicated to topological assessment for curvilinear structures, and (4) β_{match}^{err} [22] which leverages persistence barcodes to obtain a spatial matching between topological objects.

For these comparisons, we use the CHASE dataset [8] that provides 28 retinal images with vascular annotations. To illustrate the behaviour of ccDice, we

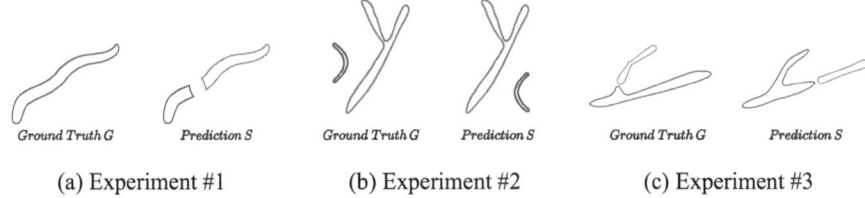

Ground Truth G   Prediction S   Ground Truth G   Prediction S   Ground Truth G   Prediction S

(a) Experiment #1   (b) Experiment #2   (c) Experiment #3

Fig. 2. Illustrations of the three proposed experiments. Green (resp. red) depicts a matched (resp. unmatched) CC. More formally, a green CC of S means that we have $\mathcal{E}(X, \varphi_{S,G}^{\lambda}(X)) \geqslant \lambda$ and a green CC of G means that we have $\mathcal{E}(X, \varphi_{G,S}^{\lambda}(X)) \geqslant \lambda$. Here λ is set to 0.5. (Color figure online)

propose three experiments where we modified artificially the annotations to generate several pairs (G, S). These experiments represent a type of error that could occur in segmentation, with a gradual escalation in the number of errors. These experiments are illustrated in Fig. 2.

Experiment #1: The goal is to study the behavior of metrics with an increasing number of disconnections in the segmentation. For each annotation, we progressively add random disconnections and compute the metrics accordingly.

Experiment #2: The goal is to study the behavior of the metrics when the number of CCs remains constant while the pixel-wise overlap of G and S progressively decreases. We start from annotations that we randomly disconnect to create G. S is then created from G by progressively removing true CCs and adding an equivalent number of false ones.

Experiment #3: The goal is to study the behavior of the metrics when the number of CCs and pixelwise overlap remain constant between S and G, but with distinct CCs. We start from an annotation, and we create S and G by randomly introducing the same number of disconnections in each, but at different locations.

4.2 Results

Figure 3 depicts the evolution of each metric in the three experiments, where the addition of errors should be reflected by a decrease of their values. In Exp. #2, the Dice and clDice decrease, since the modifications to the CCs are carried out spatially. However, they remain close to 1 (best score) in Exps. #1 and #3 as the disconnections in S do not significantly alter the pixel-wise / medial axis coherence between S and G. In Exp. #1, β_0^{err} increases, since the number of CCs between S and G diverges. However in Exps. #2 and #3, it remains close to 0 (best score) as the number of CCs remains unchanged. This shows that neither the Dice/clDice nor β_0^{err} are able to correctly assess the topological relevance of a segmentation across all scenarios. The combination of both metrics also fails in some cases, as evidenced by Exp. #3.

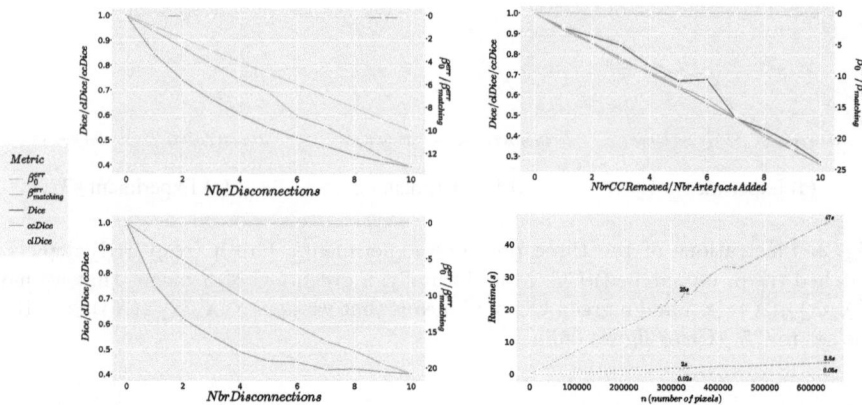

Fig. 3. Experiments #1 (upper-left), #2 (upper-right), #3 (bottom-left). Average run-times per image with respect to image size (bottom-right). The sizes of images vary from 64×64 to 850×850. Dice (purple), clDice (yellow), ccDice (red), β_0^{err} (green) and β_{match}^{err} (blue) (Color figure online)

Both ccDice and β_{match}^{err} exhibit correct behaviors in the three experiments. Their respective values decrease when the number of CCs increases (Exp. #1) or when the spatial matching between the CCs of S and G diverges (Exps. #2 and #3). However, ccDice presents various advantages. First it is a normalized metric, with values in $[0, 1]$, similarly to Dice. By contrast, β_{match}^{err} can be arbitrarily high which makes comparisons between images difficult. Second, ccDice can be computed with a low time complexity, which is never greater than quasi-linear (or even linear when $\lambda > 0.5$). Figure 3 (bottom-right) presents the average running times of the metrics with respect to the image size on the CHASE dataset. The running time of ccDice is very close to that of Dice, clDice and β_0^{err}, whereas the running time of β_{match}^{err} is significantly greater. Third, we propose a 3D implementation of ccDice[1], whereas the β_{match}^{err} is only in 2D.

5   Discussion

Analysing Holes with ccDice. ccDice allows one to compare two binary images $S, G \subseteq \Omega$ with respect to their respective "objects", i.e. the CCs $X \in \mathcal{C}[S]$ and $Y \in \mathcal{C}[G]$, which correspond to the first homology groups of S and G. It is possible to use ccDice to investigate the "holes" of S, G, which correspond to their last homology group. We can apply ccDice on the complements $\overline{S} = \Omega \setminus S$ and $\overline{G} = \Omega \setminus G$, i.e. compute ccDice($\overline{S}, \overline{G}$). It is possible to combine ccDice(S, G) and ccDice($\overline{S}, \overline{G}$) to get access to a normalized metric that evaluates a segmentation with respect to these two homology groups.

---

[1] https://github.com/PierreRouge/ccDice.

Towards ccDice Loss. One significant limitation of our method is its lack of differentiability, preventing its use as a loss function. In the following discussion, we introduce two extensions of the ccDice for fuzzy images and grey-level images, respectively. This sets the stage for future research aimed at developing a differentiable version of the ccDice.

Soft ccDice. It is possible to extend ccDice to a soft version that evaluates fuzzy images $\mathcal{S}, \mathcal{G} : \Omega \to [0,1]$, instead of binary images $S, G \subseteq \Omega$. A continous soft version of ccDice can be defined as follows:

$$\text{ccDice}_{\text{Soft}}(\mathcal{S}, \mathcal{G}) = \int_0^1 \text{ccDice}(\tau_v(\mathcal{S}), \tau_v(\mathcal{G})) dv \qquad (11)$$

where τ_v is the upper thresholding at value v function defined by $\tau_v(\mathcal{S}) = \{x \in \Omega \mid \text{S}(x) \geqslant v\}$. This soft version is compliant with ccDice. Indeed, if $\mathcal{S}, \mathcal{G} : \Omega \to [0,1]$ are binary functions, i.e. if $\mathcal{S}(x)$ and $\mathcal{G}(x) = 0$ or 1 for any $x \in \Omega$, we have $\text{ccDice}_{\text{Soft}}(\mathcal{S}, \mathcal{G}) = \text{ccDice}(\tau_1(\mathcal{S}), \tau_1(\mathcal{G}))$. This allows in particular to compare e.g. a fuzzy segmentation map $\mathcal{S} = \Omega \to [0,1]$ with a binary ground truth $G \subseteq \Omega$.

In practice, Eq. (11) is computed in a discrete way, by sampling a finite number of values in $[0,1]$. It is also possible to weight the ccDice components of this sum, leading to a computationally tractable formulation:

$$\text{ccDice}_{\text{Soft}}(\mathcal{S}, \mathcal{G}) = \sum_{v \in V} \alpha_v \cdot \text{ccDice}(\tau_v(\mathcal{S}), \tau_v(\mathcal{G})) \qquad (12)$$

where $V \subset [0,1]$ is a finite set, $\alpha_v > 0$ and $\sum_{v \in V} \alpha_v = 1$. In particular, $\text{ccDice}_{\text{Soft}}$ can be computed with a time cost $\mathcal{O}(|V| \cdot n \log n)$.

Grey-Level ccDice. More generally, one may want to define a generalization of ccDice to grey-level images without requiring to explicitly build upon persistent homology paradigm [22]. A convenient way may consist of relying on the notions of component-trees (min- and max-tree) [20] or the unifying complete tree of shapes [15], that allow to model the hierarchical structure of all the CCs (objects and holes) of the threshold sets of grey-level functions $\mathcal{S}, \mathcal{G} : \Omega \to \mathbb{R}$. In particular, these trees can be built in quasi-linear time [3] and can be involved in the design of differentiable losses [16]. These properties open the way to future developments of a grey-level version of ccDice.

6 Conclusion

In this paper, we introduced ccDice, which builds upon the notions of CCs and spatial matching to provide a metric evaluating both the topology and spatial matching of segmentations. Similarly to the Dice metric, it is a normalized, interpretable metric. It also has the notable property to generalize the Dice metric from pixels to CCs (Proposition 5), which emphasizes its theoretical soundness. Similarly to β_{match}^{err}, it takes into account topological and spatial embedding of the images in order to evaluate the relevance of a segmentation. In 2D, ccDice is as discriminant as its concurrent the β^{err}_{match} as it can assess the two first homology groups, but it is computation time is largely lower. Finally, ccDice appears as a relevant tool for topological analysis of segmentation, which opens the way to further fuzzy / grey-level extensions, a step towards developing a differentiable version of the ccDice.

Acknowledgments. This work was supported by the *Agence Nationale de la Recherche* (Grants ANR-20-CE45-0011, ANR-22-CE45-0034, ANR-22-CE45-0018 and ANR-23-CE45-0015).

Disclosure of Interests. The authors have no competing interests to declare that are relevant to the content of this article.

References

1. Bauer, U., Lesnick, M.: Induced matchings and the algebraic stability of persistence barcodes. J. Comput. Geom. **6**, 162–191 (2015)
2. Byrne, N., et al.: A persistent homology-based topological loss for CNN-based multiclass segmentation of CMR. IEEE Trans. Med. Imaging **42**, 3–14 (2022)
3. Carlinet, E., Géraud, T.: A comparative review of component tree computation algorithms. IEEE Trans. Image Process. **23**, 3885–3895 (2014)
4. Clough, J.R., et al.: A topological loss function for deep-learning based image segmentation using persistent homology. IEEE Trans. Pattern Anal. **44**, 8766–8778 (2020)
5. Couprie, M., Bertrand, G.: New characterizations of simple points in 2D, 3D, and 4D discrete spaces. IEEE Trans. Pattern Anal. **31**, 637–648 (2009)
6. Dice, L.R.: Measures of the amount of ecologic association between species. Ecology **26**, 297–302 (1945)
7. Edelsbrunner, H., Harrer, J.: Persistent homology - a survey. Contemp. Math. **453**, 257–282 (2008)
8. Fraz, M.M., et al.: An ensemble classification-based approach applied to retinal blood vessel segmentation. IEEE Trans. Bio. Med. Eng. **59**, 2538–2548 (2012)
9. Gupta, S., et al.: Topology-aware uncertainty for image segmentation. In: NeurIPS, Procs. (2024)
10. Hu, X., et al.: Topology-preserving deep image segmentation. In: NeurIPS, Procs. (2019)
11. Hu, X., et al.: Topology-aware segmentation using discrete Morse theory. In: ICLR, Procs. (2021)
12. Kovalevsky, V.A.: Finite topology as applied to image analysis. Comput. Vision. Graph. **46**, 141–161 (1989)
13. Mazo, L., et al.: Paths, homotopy and reduction in digital images. Acta Appl. Math. **113**, 167–193 (2011)
14. Mazo, L., et al.: Digital imaging: a unified topological framework. J. Math. Imaging Vis. **44**, 19–37 (2012)
15. Passat, N., Mendes Forte, J., Kenmochi, Y.: Morphological hierarchies: a unifying framework with new trees. J. Math. Imaging Vis. **65**, 718–753 (2023)
16. Perret, B., Cousty, J.: Component tree loss function: Definition and optimization. In: DGMM, Procs., pp. 248–260 (2022)

17. Rosenfeld, A.: Adjacency in digital pictures. Inform. Control **26**, 24–33 (1974)
18. Rosenfeld, A.: Digital topology. Am. Math. Mon. **86**, 621–630 (1979)
19. Saha, P.K., Strand, R., Borgefors, G.: Digital topology and geometry in medical imaging: a survey. IEEE Trans. Med. Imaging **34**, 1940–1964 (2015)
20. Salembier, P., Oliveras, A., Garrido, L.: Anti-extensive connected operators for image and sequence processing. IEEE Trans. Image Process. **7**, 555–570 (1998)
21. Shit, S., et al.: clDice–A novel topology-preserving loss function for tubular structure segmentation. In: CVPR, Procs., pp. 16560–16569 (2021)
22. Stucki, N., et al.: Topologically faithful image segmentation via induced matching of persistence barcodes. In: ICML, Procs., pp. 32698–32727 (2023)

TopOC: Topological Deep Learning for Ovarian and Breast Cancer Diagnosis

Saba Fatema[1], Brighton Nuwagira[1], Sayoni Chakraborty[1], Reyhan Gedik[2], and Baris Coskunuzer[1(✉)]

[1] Department of Mathematical Sciences, The University of Texas at Dallas, Richardson, TX 75080, USA
{saba.fatema,brighton.nuwagira,sayoni.chakraborty, coskunuz}@utdallas.edu
[2] MGH - Pathology Department, Harvard Medical School, Boston, MA 02114, USA
rgedik@mgh.harvard.edu

Abstract. Microscopic examination of slides prepared from tissue samples is the primary tool for detecting and classifying cancerous lesions, a process that is time-consuming and requires the expertise of experienced pathologists. Recent advances in deep learning methods hold significant potential to enhance medical diagnostics and treatment planning by improving accuracy, reproducibility, and speed, thereby reducing clinicians' workloads and turnaround times. However, the necessity for vast amounts of labeled data to train these models remains a major obstacle to the development of effective clinical decision support systems.

In this paper, we propose the integration of topological deep learning methods to enhance the accuracy and robustness of existing histopathological image analysis models. Topological data analysis (TDA) offers a unique approach by extracting essential information through the evaluation of topological patterns across different color channels. While deep learning methods capture local information from images, TDA features provide complementary global features. Our experiments on publicly available histopathological datasets demonstrate that the inclusion of topological features significantly improves the differentiation of tumor types in ovarian and breast cancers.

Keywords: Ovarian Cancer Diagnosis · Breast Cancer Diagnosis · Histopathology · Cubical Persistence · Topological Data Analysis

1 Introduction

Ovarian and breast cancers rank among the most common and lethal malignancies impacting women globally. The cornerstone of detecting and classifying these cancers is histopathology, which involves the microscopic examination of tissue samples and cellular collections. This meticulous process is vital for accurate diagnosis and treatment planning but is often time-consuming and heavily dependent on the expertise of seasoned pathologists. As the demand for precise and rapid diagnostics increases,

Supplementary Information The online version contains supplementary material available at https://doi.org/10.1007/978-3-031-73967-5_3.

the constraints of traditional methods become more evident, underscoring the need for advanced technologies to support and enhance pathological workflows.

Recent advancements in deep learning have shown great promise in the field of medical imaging, offering the potential to revolutionize diagnostics by providing improved accuracy, reproducibility, and efficiency. Deep learning algorithms excel at identifying complex patterns within images, making them well-suited for the detailed analysis required in histopathology. However, the effectiveness of these models is often hindered by the substantial requirement for large, labeled datasets, which can be difficult and costly to obtain.

To address these challenges, we explore the integration of topological data analysis (TDA) tools with deep learning methods for histopathological image analysis. TDA provides a novel approach by focusing on the extraction and evaluation of topological patterns within images, capturing global features that complement the local information typically identified by deep learning algorithms. This synergy between local and global feature extraction can enhance the robustness and accuracy of cancer detection models.

In this paper, we present a method that incorporates topological deep learning techniques into the analysis of histopathological images. We demonstrate the effectiveness of this approach using publicly available benchmark datasets for ovarian and breast cancer, showing that our method significantly improves the differentiation of tumor types. Our results suggest that the integration of TDA with deep learning not only enhances model performance but also has the potential to streamline diagnostic workflows, ultimately benefiting both clinicians and patients.

Our contributions can be summarized as follows:

◇ We introduce two robust topological machine learning models for the analysis of histopathological images.

◇ Using cubical persistence, we extract unique topological fingerprints by examining the evolution of topological patterns in the images across each color channel.

◇ Our first model, TopOC-1, applies standard machine learning models to topological vectors, making it straightforward, interpretable, and computationally feasible without the need for data augmentation.

◇ Our second model, TopOC-CNN, integrates topological features with existing pre-trained models. We observe consistent performance improvements with the inclusion of these topological features.

◇ Both models deliver outstanding performance on ovarian and breast cancer histopathological image datasets, surpassing state-of-the-art models.

2   Related Work

2.1   Machine Learning Methods for Ovarian and Breast Cancer Diagnosis

Deep learning (DL) techniques show significant potential in enhancing ovarian cancer diagnosis through histopathological image analysis, identifying malignant patterns, and aiding clinicians in decision-making [29]. [15] presents DL model for ovarian carcinoma histotype classification comparable to expert pathologists. Various CNN models like YOLO, DenseNet, GoogleNet(V3), ResNet, etc., have demonstrated competitive accuracy with medical professionals using ultrasound images [24,39].

A hybrid DL model using multi-modal data (gene and histopathology images) is proposed by [16] to address the limitations of single-modal approaches in representing ovarian cancer characteristics. An approach for better ovarian cancer risk stratification by integrating histopathological, radiologic, and clinicogenomic data is demonstrated in [6]. A review in [7] assesses AI in ovarian cancer pathology, identifies clinical gaps, and recommends improvements for future adoption.

DL methods are also employed in various recent studies to detect breast cancer. In [19], authors introduce a model for detecting breast cancer in mammograms of varying densities, using low-variance elimination, univariate selection, and recursive feature elimination as feature selection modules. [41] proposed a multimodal machine learning model for breast cancer detection that combines mammograms and ultrasound data. For a thorough review of the breast cancer histopathological image analysis methods, see [10,42], where the authors review the recent models based on artificial neural networks, categorizing them into classical and deep neural network approaches.

2.2 TDA in Medical Image Analysis

Persistent homology (PH) has been highly effective for pattern recognition in image and shape analysis over the past two decades. In medical image analysis, PH has produced powerful results in cell development [20], tumor detection [13,37], histopathology [25], brain functionality analysis [9], fMRI data [27], and genomic data [26]. For a thorough review of TDA methods in biomedicine, see the excellent survey [31]. For a collection of TDA applications in various domains, see the TDA Applications Library [17, 18].

Recently, there has been a surge of interest in topological deep learning within the machine learning community, showing great potential to augment existing deep learning methodologies [23,43]. The effective utilization of topological features has notably enhanced CNN models, particularly in tasks such as image segmentation [19,30,36]. Moreover, there is growing recognition of the pivotal role of topological features in diagnostic tasks across various medical domains [32,40]. To our knowledge, our work presents the first application of TDA methods for the detection of ovarian cancer.

3   Methodology

Our methodology consists of two primary steps. First, we extract the topological feature vectors from the images. Then, we apply these topological features directly to machine learning models, resulting in a fast and interpretable model called *TopOC-1*. Next, we integrate these vectors with existing deep learning models for cancer diagnosis, resulting in *TopOC-CNN*.

3.1   Topological Vectors for Images

Persistent homology (PH) serves as a robust mathematical tool within topological data analysis (TDA) for exploring the intricate shape and structure inherent in complex datasets. Its fundamental concept involves systematically examining the development of various hidden patterns within the data at different resolutions [11]. PH demonstrates remarkable effectiveness in extracting features from diverse data formats such as

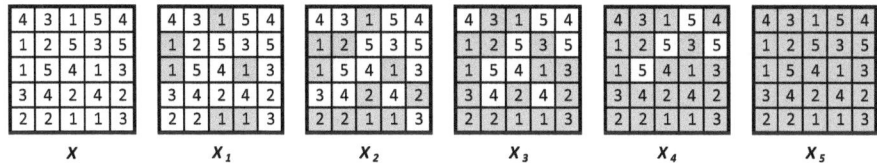

Fig. 1. For the 5×5 image $\mathcal{X}$ with the given pixel values, **the sublevel filtration** is the sequence of binary images $\mathcal{X}_1 \subset \mathcal{X}_2 \subset \mathcal{X}_3 \subset \mathcal{X}_4 \subset \mathcal{X}_5$.

point clouds and networks. However, our paper focuses exclusively on its application in image analysis, specifically emphasizing *cubical persistence*, which is a specific version of PH. While we aim to describe the construction of PH in accessible terms for non-specialists, those seeking deeper insights can refer to the comprehensive textbook [14]. Essentially, PH can be summarized as a three-step procedure as follows.

1. **Filtration**: Inducing a sequence of nested topological spaces from the data.
2. **Persistence Diagrams**: Recording the evolution of topological changes within this sequence.
3. **Vectorization**: Converting these persistence diagrams into vectors to be utilized in machine learning models.

Step 1 - Constructing Filtrations. Since PH essentially functions as a mechanism for monitoring the progression of topological characteristics within a sequence of simplicial complexes, constructing this sequence stands out as a crucial step. In image analysis, the common approach is to generate a nested sequence of binary images, also known as *cubical complexes*. To achieve this from a given color (or grayscale) image $\mathcal{X}$ (with dimensions $r \times s$), one needs to select a specific color channel (e.g., red, blue, green, or grayscale). The color values γ_{ij} of individual pixels $\Delta_{ij} \subset \mathcal{X}$ are then utilized. Specifically, for a sequence of color values ($0 = t_1 < t_2 < \cdots < t_N = 255$), a nested sequence of binary images $\mathcal{X}_1 \subset \mathcal{X}_2 \subset \cdots \subset \mathcal{X}_N$ is obtained, where $\mathcal{X}_n = \{\Delta_{ij} \subset \mathcal{X} \mid \gamma_{ij} \leq t_n\}$ (See Fig. 1).

In particular, this involves starting with a blank $r \times s$ image and progressively activating (coloring black) pixels as their grayscale values reach the specified threshold t_n. This process, known as *sublevel filtration*, is conducted on $\mathcal{X}$ relative to a designated function (in this instance, grayscale). Alternatively, one can activate pixels in descending order, referred to as *superlevel filtration*. In other words, let $\mathbf{Y}_n = \{\Delta_{ij} \subset \mathcal{X} \mid \gamma_{ij} \geq s_n\}$ for ($255 = s_1 > s_2 > \cdots > s_M = 0$), and $\mathbf{Y}_1 \subset \mathbf{Y}_2 \subset \cdots \subset \mathbf{Y}_M$ is called superlevel filtration.

Step 2 - Persistence Diagrams. PH traces the development of topological characteristics across the filtration sequence $\{\mathcal{X}_n\}$ and presents it through a *persistence diagram* (PD). Specifically, if a topological feature σ emerges in $\mathcal{X}_m$ and disappears in $\mathcal{X}_n$ with $1 \leq m < n \leq N$, the thresholds t_m and t_n are denoted as the *birth time* b_σ and *death time* d_σ of σ, respectively ($b_\sigma = t_m$ and $d_\sigma = t_n$). Therefore, PD contains all such 2-tuples $\mathrm{PD}_k(\mathcal{X}) = \{(b_\sigma, d_\sigma)\}$ where k represents the dimension of the topological features. The interval $d_\sigma - b_\sigma$ is termed as the *lifespan* of σ. Formally, the kth persistence

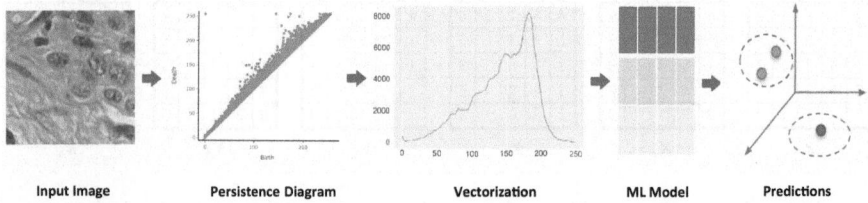

| Input Image | Persistence Diagram | Vectorization | ML Model | Predictions |

Fig. 2. TopOC-1 Model. We first generate persistence diagrams for any input image, utilizing grayscale values. Next, we derive our topological feature vectors, represented as Betti functions. These vectors are then inputted into a simple ML classifier to produce classification results.

diagram can be defined as $\mathrm{PD}_k(\mathcal{X}) = \{(b_\sigma, d_\sigma) \mid \sigma \in H_k(\mathcal{X}_n) \text{ for } b_\sigma \leq t_n < d_\sigma\}$, where $H_k(\mathcal{X}_n)$ denotes the kth homology group of the cubical complex $\mathcal{X}_n$. Thus, $\mathrm{PD}_k(\mathcal{X})$ contains 2-tuples indicating the birth and death times of k-dimensional voids $\{\sigma\}$ (such as connected components, holes, and cavities) in the filtration sequence $\{\mathcal{X}_n\}$. For instance, for $\mathcal{X}$ in Fig. 1, $\mathrm{PD}_0(\mathcal{X}) = \{(1, \infty), (1, 2), (1, 2), (1, 3)\}$ represents the connected components, while $\mathrm{PD}_1(\mathcal{X}) = \{(2, 4), (3, 5), (4, 5)\}$ illustrates the holes in the corresponding binary images in Fig. 1.

Step 3 - Vectorization. Persistence Diagrams (PDs), which consist of collections of 2-tuples, are not practical for use with ML tools. Instead, a common approach is to convert PD information into a vector or function, a process known as *vectorization* [2].

One commonly used function for this purpose is the *Betti function*, which tracks the number of *alive* topological features at each threshold. Specifically, the Betti function is a step function where $\beta_0(t_n)$ represents the count of connected components in the binary image $\mathcal{X}_n$, and $\beta_1(t_n)$ indicates the number of holes (loops) in $\mathcal{X}_n$. In the context of ML, Betti functions are typically represented as vectors $\vec{\beta_k}$ of size N with entries $\beta_k(t_n)$ for $1 \leq n \leq N$, defined as $\vec{\beta_k}(\mathcal{X}) = [\beta_k(t_1) \ \beta_k(t_2) \ \dots \ \beta_k(t_N)]$.

For example, in Fig. 1, we can observe $\vec{\beta_0}(\mathcal{X}) = [4 \ 2 \ 1 \ 1 \ 1]$, indicating the count of connected components in the binary images $\{\mathcal{X}_i\}$, while $\vec{\beta_1}(\mathcal{X}) = [0 \ 1 \ 2 \ 2 \ 0]$ represents the counts of holes in $\{\mathcal{X}_i\}$. Notably, $\beta_0(1) = 4$ signifies the count of components in $\mathcal{X}_1$, and $\beta_1(3) = 2$ denotes the count of holes (loops) in $\mathcal{X}_3$.

There exist various other methods to convert PDs into a vector, such as persistence images [1], persistence landscapes [8], silhouettes [12], and kernel methods [2]. However, in this paper, we primarily utilize Betti functions due to their computational efficiency and ease of interpretation.

3.2   TopOC Models

Once we have obtained the topological vectors, the next step involves applying ML tools to analyze the effectiveness of these features. Essentially, each medical image is represented as a 200-dimensional vector (300-dimensional for 3D images), signifying its embedding in a latent space. The classification task treats these embeddings as a point cloud and aims to identify distinct clusters corresponding to different classes.

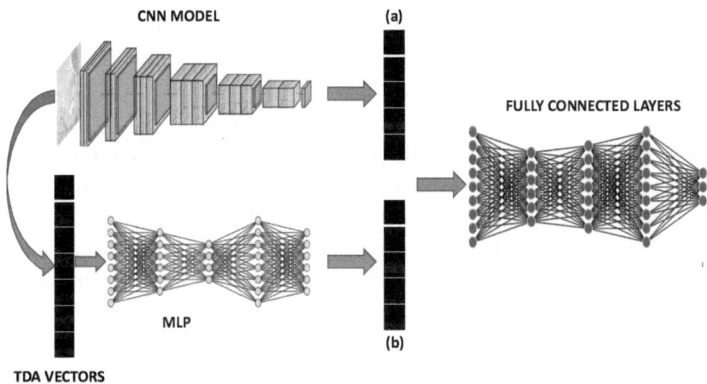

Fig. 3. TopOC-CNN Model. In this model, we integrate topological features of images with convolutional vectors from a CNN backbone, followed by a fully connected layer. For the prediction head, we concatenate (a) a 128-dimensional CNN vector and (b) a topological vector, with dimension options of 256, 128, 64, or 0 (Vanilla-CNN).

To evaluate the effectiveness of these topological vectors, we employed two different *Topological Ovarian Cancer (TopOC)* models.

TopOC-1 model. In our basic model, the TopOC-1 model (Fig. 2), we directly used ML classifier on topological embeddings of images to test the performance of topological vectors. The main idea is to use the procedure described in Sect. 3.1, for each image, we define a sublevel filtration with 50 thresholds to span $[0, 255]$ color interval for each color. For an RGB image, we use Red, Green, Blue, and Grayscale (average of RGB color values) color values for each pixel, and we obtain four different sublevel filtrations one for each color. Next, we obtain the corresponding persistence diagrams, and by applying Betti vectorization, we obtain topological vectors for each image. Each color channel induces 50 dimensional $\beta_0(\mathcal{X})$ and $\beta_1(\mathcal{X})$. By concatenating these vectors, we obtain a 400-dimensional vector for each image. To eliminate correlations, we use dimension reduction by using feature selection methods. In our first model, TopOC-1, we simply apply standard ML classifiers on these vectors. Next, in our more sophisticated method, we integrate these vectors with deep learning methods.

TopOC-CNN Model. In our second model, TopOC-CNN (Fig. 3), we integrated topological features with pre-trained CNN models. This integration is based on the concept that topological features capture global patterns within the image, while convolutional vectors focus on more localized aspects. Combining these two distinct sets of features is expected to enhance the model's overall performance. The TopOC-CNN model is meticulously crafted by selecting a base model from a range of state-of-the-art CNN models, including EfficientNetB0, DenseNet121, and VGG16. These models were chosen for their robust feature extraction capabilities, leveraging pre-trained weights from extensive datasets like ImageNet. The layers of these pre-trained CNN models remain frozen, and their capabilities are augmented by adding extra layers to enhance their ability to capture intricate patterns and features. The CNN component of our model starts with an input layer that accepts images with a shape of $(224, 224, 3)$. This input is processed by the base model, whose output is then fed into an additional convolutional layer containing 256 filters, each with a 3x3 kernel size and ReLU activation function. Subsequently, a max pooling layer with a 2x2 pool size is applied to downsample the feature maps. These feature maps are then flattened into a single vector, which is further refined by a dense layer with 128 neurons and a ReLU activation function. Simultaneously, a Multilayer Perceptron (MLP) is defined with three dense layers, each utilizing the Rectified Linear Unit (ReLU) activation function. The MLP layers consist of 256 neurons each, with the number of neurons in the last layer varying for different settings: 256, 128, 64, and 0 neurons. The outputs of both the augmented CNN layers (128 neurons) and the MLP (256, 128, 64, or 0 neurons) are concatenated and passed through additional dense layers (256 neurons, 128 neurons, 128 neurons), all with ReLU activation functions, ultimately culminating in the output layer.

4   Experiments

4.1   Experimental Setup

Datasets.   To validate the performance of our models, we used two publicly available histopathology datasets. The first dataset, UBC-OCEAN, was provided in the UBC ovarian cancer subtype classification and outlier detection (UBC-OCEAN) challenge hosted by the University of British Columbia [3, 15]. The dataset is accessible at the Kaggle link [4] and provides 34K tiles of 224×224 size obtained from 975 whole slide images (WSI) and tissue microarrays (TMA). The task is to classify five subtypes of ovarian cancer, namely clear cell ovarian carcinoma (CC), endometrioid (EC), high-grade serous carcinoma (HGSC), low-grade serous carcinoma (LGSC), and mucinous carcinoma (MC). We used the pre-defined training and test split (91:9) provided by the challenge, i.e., $31, 203$ training and $3, 082$ test images. Our second dataset, BREAKHIS, is the Breast Cancer Histopathology dataset [33], consisting of 7909 samples with two classes, Benign and Malignant. These samples are provided at 700×460 size with different magnification factors such as 40x, 100x, 200x, and 400x [34, 35]. For this dataset, we used a commonly employed 70:30 training-test split [5].

Topological Vectors.   As described in Sect. 3.1, for each image, by employing a filtration for each color channel (R, G, B, and G), we obtain four filtration. We employed 50-thresholds, and each color channel, produces 50-dimensional $\vec{\beta_0}$ and $\vec{\beta_1}$ vectors (See Fig. 4). Next, by concatenating them, we got a 400-dimensional topological vector for each image.

In Appendix A, we give further experimental details, including *Model Hyperparameters, Extra Performance Metrics,* and *Computational Complexity and Runtime.*

4.2   Results

UBC-OCEAN is a recent challenge that concluded a few months ago, offering a $50K prize. The performance metric was balanced accuracy, and the winning entry achieved 66% balanced accuracy for this five-class classification task [4]. Both of our models surpassed this benchmark. TopOC-1, using 200 features, achieved 66.13% balanced accuracy as shown in Table 1. Our TopOC-CNN model, employing a DenseNet121

Table 1. TopOC-1. Performances of our TopOC-1 model with different numbers of topological features.

Models	UBC-OCEAN			BREAKHIS-40x		
	B. Acc.	Acc.	AUC	Acc.	Sen.	Spec.
50 Features	61.14	67.32	89.07	87.81	94.51	74.24
100 Features	63.45	69.92	90.94	89.32	96.51	74.75
200 Features	**66.13**	**72.45**	91.74	**90.82**	**97.76**	**76.77**
400 Features	65.31	71.51	**91.75**	89.98	97.76	74.24

backbone and 128-dimensional topological vectors, reached 67.15% balanced accuracy. Notably, we utilized only pre-trained models without any data augmentation, whereas many competitors relied on customized CNN models with extensive data augmentation and other optimizations.

Table 2. Balanced Accuracy, Accuracy, and AUC performances for TopOC-CNN models with different backbones and varying topological MLP final output layer settings (64, 128, 256) on the multiclass classification of UBC-OCEAN dataset.

Models	DenseNet121			EfficientNetB0			VGG16		
	B. Acc.	Acc.	AUC	B. Acc.	Acc.	AUC	B. Acc.	Acc.	AUC
Vanilla-CNN	65.10	69.08	89.42	56.08	63.82	87.04	56.45	61.55	84.02
CNN+64Top	64.62	68.59	89.37	59.85	64.34	86.71	56.50	61.26	83.53
CNN+128Top	**67.15**	69.34	**92.08**	59.96	64.05	86.48	55.00	62.20	**84.33**
CNN+256Top	66.36	**69.99**	89.62	**62.67**	**67.98**	**89.14**	**57.63**	**63.24**	84.24
Improvement	2.05	0.91	2.66	6.59	4.16	2.10	1.18	1.69	0.31

Table 3. Performance comparison of Vanilla-CNN and TopOC-CNN models with different backbones for BREAKHIS dataset.

Mag.	Models	DenseNet121			EfficientNetB0			VGG16		
		Acc	Sens.	Spec	Acc	Sens	Spec	Acc	Sens	Spec
40x	Vanilla-CNN	92.99	91.72	88.30	84.81	84.46	83.51	88.81	87.38	83.51
	CNN+256Top	94.82	94.06	92.02	93.16	91.12	85.64	90.48	89.31	86.17
	Improvement	1.83	2.34	3.72	8.35	6.66	2.13	1.67	1.93	2.66
100x	Vanilla-CNN	90.40	88.18	82.38	89.12	85.68	76.68	87.84	84.18	74.61
	CNN+256Top	92.48	90.98	87.05	92.16	91.03	88.08	92.32	90.57	86.01
	Improvement	2.08	2.80	4.67	3.04	5.35	11.4	4.48	6.39	11.4
200x	Vanilla-CNN	91.72	90.17	86.10	87.58	84.08	74.86	87.58	85.40	79.68
	CNN+256Top	93.71	92.79	90.37	93.71	91.76	86.63	90.07	88.23	83.42
	Improvement	1.99	2.62	4.27	6.13	7.68	11.77	2.49	2.83	3.74
400x	Vanilla-CNN	91.03	88.16	80.11	85.35	81.74	71.59	85.53	81.88	71.59
	CNN+256Top	93.22	92.17	89.20	90.29	87.18	78.41	87.73	84.84	76.70
	Improvement	2.19	4.01	9.09	4.94	5.44	6.82	2.20	2.96	5.11

In Table 2, we present the TopOC-CNN results for three backbones (DenseNet121, EfficientNetB0, VGG16) and four variations of each. Specifically, in Fig. 3, the CNN vector (a) is fixed at 128-dimension across all models, while the topological vector (b) (MLP output) varies among 0 (Vanilla-CNN), 64, 128, and 256-dimension. Our findings indicate that incorporating topological vectors consistently enhances CNN performance, up to 6.59% improvements in balanced accuracy.

For the BREAKHIS dataset, both of our models also deliver highly competitive results in breast cancer diagnosis (Table 4). Similarly, our topological vectors enhance CNN model performance by up to 8.35% in accuracy (Table 3).

We provide our model's *Visualization and Interpretability* discussion in Appendix B.

Table 4. Performance comparison of TopOC model with deep learning models on binary classification of BREAKHIS dataset.

Method	40x			100x			200x			400x		
	Acc.	Sens.	Spec.	Acc.	Sens.	Spec.	Acc.	Sens.	Spec.	Acc.	Sens.	Spec.
AlexNet [35]	81.52	75.64	87.40	81.28	78.16	84.40	83.54	79.16	87.91	81.10	76.28	85.90
BkNet [38]	85.61	84.42	86.80	86.23	87.19	85.26	85.37	80.01	90.74	84.43	80.00	88.87
CapsNet [28]	86.95	86.29	87.61	89.13	88.30	89.96	88.75	86.22	91.28	88.04	87.56	88.51
BkCapsNet [38]	92.71	92.15	**93.27**	**94.52**	95.16	**93.87**	**94.03**	94.31	**93.75**	**93.54**	94.06	**93.03**
TopOC-1	90.82	**97.76**	76.77	91.68	**96.00**	82.50	**96.53**	80.00	88.64	**96.02**	75.25	
TopOC-CNN	**94.82**	94.06	92.02	92.48	91.03	88.08	93.71	92.79	90.37	93.22	92.17	89.20

5   Conclusion

In conclusion, our study underscores the substantial promise of integrating topological machine learning methods with existing deep learning techniques for histopathological image analysis in ovarian and breast cancers. The developed models, TopOC-1 and TopOC-CNN, exhibit enhanced accuracy and efficiency in cancer detection by incorporating topological features into the deep learning framework. Moving along, we aim to focus on expanding model validation across diverse datasets and cancer types, refining topological feature extraction techniques, and incorporating these methods into real-time clinical diagnostic tools. Addressing the challenge of obtaining large labeled datasets through collaboration and advanced learning techniques will further optimize these models, ultimately advancing more accurate and accessible cancer diagnostics.

Acknowledgments. This work was partially supported by the National Science Foundation under grants DMS-2202584, 2229417, and DMS-2220613 and by Simons Foundation under grant # 579977. The authors acknowledge the Texas Advanced Computing Center (TACC) at UT Austin for computational resources which contributed to the research results reported within this paper.

References

1. Adams, H., et al.: Persistence images: a stable vector representation of persistent homology. J. Mach. Learn. Res. **18**(1), 218–252 (2017)
2. Ali, D., et al.: A survey of vectorization methods in topological data analysis. IEEE Trans. Pattern Anal. Mach. Intell. **45**(12), 14069–14080 (2023)
3. Asadi-Aghbolaghi, M., et al.: Machine learning-driven histotype diagnosis of ovarian carcinoma: Insights from the ocean AI challenge. medRxiv, pp. 2024–04 (2024)
4. Bashashati, A., et al.: UBC ovarian cancer subtype classification and outlier detection (UBC-OCEAN) (2023). https://kaggle.com/competitions/UBC-OCEAN
5. Benhammou, Y., Achchab, B., Herrera, F., Tabik, S.: BreakHis based breast cancer automatic diagnosis using deep learning. Neurocomputing **375**, 9–24 (2020)
6. Boehm, K.M., et al.: Multimodal data integration using machine learning improves risk stratification of high-grade serous ovarian cancer. Nat. Cancer **3**(6), 723–733 (2022)
7. Breen, J., et al.: Artificial intelligence in ovarian cancer histopathology: a systematic review. NPJ Precis. Oncol. **7**(1), 83 (2023)
8. Bubenik, P., Dłotko, P.: A persistence landscapes toolbox for topological statistics. J. Symb. Comput. **78**, 91–114 (2017)
9. Caputi, L., Pidnebesna, A., Hlinka, J.: Promises and pitfalls of topological data analysis for brain connectivity analysis. Neuroimage **238**, 118245 (2021)
10. Chan, R.C., To, C.K.C., Cheng, K.C.T., Yoshikazu, T., Yan, L.L.A., Tse, G.M.: Artificial intelligence in breast cancer histopathology. Histopathology **82**(1), 198–210 (2023)
11. Chazal, F., Michel, B.: An introduction to topological data analysis: fundamental and practical aspects for data scientists. Front. Artif. Intell. **4**, 667963 (2021)
12. Chazal, F., Fasy, B.T., Lecci, F., Rinaldo, A., Wasserman, L.: Stochastic convergence of persistence landscapes and silhouettes. In: SoCG, pp. 474–483 (2014)
13. Crawford, L., Monod, A., Chen, A.X., Mukherjee, S., Rabadán, R.: Predicting clinical outcomes in glioblastoma: an application of topological and functional data analysis. J. Am. Stat. Assoc. **115**(531), 1139–1150 (2020)
14. Dey, T.K., Wang, Y.: Computational Topology for Data Analysis. Cambridge University Press (2022)
15. Farahani, H., et al.: Deep learning-based histotype diagnosis of ovarian carcinoma whole-slide pathology images. Mod. Pathol. **35**(12), 1983–1990 (2022)
16. Ghoniem, R.M., Algarni, A.D., Refky, B., Ewees, A.A.: Multi-modal evolutionary deep learning model for ovarian cancer diagnosis. Symmetry **13**(4), 643 (2021)
17. Giunti, B.: TDA applications library (2022). https://www.zotero.org/groups/2425412/tda-applications/library
18. Giunti, B., Lazovskis, J., Rieck, B.: DONUT: Database of original & non-theoretical uses of topology (2022). https://donut.topology.rocks
19. Gupta, S., et al.: Learning topological interactions for multi-class medical image segmentation. In: Avidan, S., Brostow, G., Cissé, M., Farinella, G.M., Hassner, T. (eds.) ECCV 2022. LNCS, vol. 13689, pp. 701–718. Springer, Cham (2022). https://doi.org/10.1007/978-3-031-19818-2_40
20. McGuirl, M.R., Volkening, A., Sandstede, B.: Topological data analysis of zebrafish patterns. Proc. Nat. Acad. Sci. **117**(10), 5113–5124 (2020)
21. Milosavljević, N., Morozov, D., Skraba, P.: Zigzag persistent homology in matrix multiplication time. In: SoCG, pp. 216–225 (2011)
22. Otter, N., Porter, M.A., Tillmann, U., Grindrod, P., Harrington, H.A.: A roadmap for the computation of persistent homology. EPJ Data Sci. **6**, 1–38 (2017)

23. Papamarkou, T., Birdal, T., Bronstein, M., Carlsson, G., et al.: Position paper: challenges and opportunities in topological deep learning. arXiv preprint arXiv:2402.08871 (2024)
24. Pham, T.L., Le, V.H.: Ovarian tumors detection and classification from ultrasound images based on YOLOv8. J. Adv. Inform. Technol. **15**(2), 264–275 (2024)
25. Qaiser, T., et al.: Fast and accurate tumor segmentation of histology images using persistent homology and deep convolutional features. Med. Image Anal. **55**, 1–14 (2019)
26. Rabadán, R., Blumberg, A.J.: Topological data analysis for genomics and evolution: topology in biology. Cambridge University Press (2019)
27. Rieck, B., et al.: Uncovering the topology of time-varying FMRI data using cubical persistence. NeurIPS **33**, 6900–6912 (2020)
28. Sabour, S., Frosst, N., Hinton, G.E.: Dynamic routing between capsules. In: Advances in Neural Information Processing Systems, vol. 30 (2017)
29. Sadeghi, M.H., et al.: Deep learning in ovarian cancer diagnosis: a comprehensive review of various imaging modalities. Pol. J. Radiol. **89**, e30 (2024)
30. Santhirasekaram, A., Winkler, M., Rockall, A., Glocker, B.: Topology preserving compositionality for robust medical image segmentation. In: CVPR, pp. 543–552 (2023)
31. Skaf, Y., Laubenbacher, R.: Topological data analysis in biomedicine: a review. J. Biomed. Inform. **130**, 104082 (2022)
32. Somasundaram, E., et al.: Persistent homology of tumor CT scans is associated with survival in lung cancer. Med. Phys. **48**(11), 7043–7051 (2021)
33. Spanhol, F., Oliveira, L.S., Petitjean, C., Heutte, L.: Breast cancer histopathological database (BreakHis) (2024). https://web.inf.ufpr.br/vri/databases/breast-cancer-histopathological-database-breakhis/
34. Spanhol, F.A., Oliveira, L.S., et al.: Deep features for breast cancer histopathological image classification. In: 2017 IEEE SMC, pp. 1868–1873. IEEE (2017)
35. Spanhol, F.A., Oliveira, L.S., Petitjean, C., Heutte, L.: Breast cancer histopathological image classification using CNNs. In: IJCNN, pp. 2560–2567. IEEE (2016)
36. Stucki, N., Paetzold, J.C., Shit, S., Menze, B., Bauer, U.: Topologically faithful image segmentation via induced matching of persistence barcodes. In: ICML. PMLR (2023)
37. Wang, F., Kapse, S., Liu, S., Prasanna, P., Chen, C.: TopoTxR: a topological biomarker for predicting treatment response in breast cancer. In: Feragen, A., Sommer, S., Schnabel, J., Nielsen, M. (eds.) IPMI 2021. LNCS, vol. 12729, pp. 386–397. Springer, Cham (2021). https://doi.org/10.1007/978-3-030-78191-0_30
38. Wang, P., et al.: Automatic classification of breast cancer histopathological images based on deep feature fusion. Biomed. Signal Process. Control **65**, 102341 (2021)
39. Wu, C., Wang, Y., Wang, F.: Deep learning for ovarian tumor classification with ultrasound images. In: Hong, R., Cheng, W.-H., Yamasaki, T., Wang, M., Ngo, C.-W. (eds.) PCM 2018. LNCS, vol. 11166, pp. 395–406. Springer, Cham (2018). https://doi.org/10.1007/978-3-030-00764-5_36
40. Yadav, A., Ahmed, F., Daescu, O., Gedik, R., Coskunuzer, B.: Histopathological cancer detection with topological signatures. In: IEEE BIBM, pp. 1610–1619. IEEE (2023)
41. Yadav, R.K., Singh, P., Kashtriya, P.: Diagnosis of breast cancer using machine learning techniques-a survey. Procedia Comput. Sci. **218**, 1434–1443 (2023)
42. Zhou, X., et al.: A comprehensive review for breast histopathology image analysis using classical and deep neural networks. IEEE Access **8**, 90931–90956 (2020)
43. Zia, A., Khamis, A., Nichols, J., Hayder, Z., Rolland, V., Petersson, L.: Topological deep learning: a review of an emerging paradigm. arXiv preprint arXiv:2302.03836 (2023)

Analyzing Brain Tumor Connectomics Using Graphs and Persistent Homology

Debanjali Bhattacharya$^{(\boxtimes)}$, Ninad Aithal, Manish Jayswal, and Neelam Sinha

Center for Brain Research, Indian Institute of Science, Bangalore, India
debanjali@cbr-iisc.ac.in

Abstract. Recent advances in molecular and genetic research have identified a diverse range of brain tumor sub-types, shedding light on differences in their molecular mechanisms, heterogeneity, and origins. The present study performs whole-brain connectome analysis using diffusion-weighted images. To achieve this, both graph theory and persistent homology-a prominent approach in topological data analysis are employed in order to quantify changes in the structural connectivity of the whole-brain connectome in subjects with brain tumors. Probabilistic tractography is used to map the number of streamlines connecting 84 distinct brain regions, as delineated by the Desikan-Killiany atlas from FreeSurfer. These streamline mappings form the connectome matrix, on which persistent homology based analysis and graph theoretical analysis are executed to evaluate the discriminatory power between tumor sub-types that include meningioma and glioma. A detailed statistical analysis is conducted on persistent homology-derived topological features and graphical features to identify the brain regions where differences between study groups are statistically significant ($p < 0.05$). For classification purpose, graph-based local features are utilized, achieving a highest accuracy of 88%. In classifying tumor sub-types, an accuracy of 80% is attained. The findings obtained from this study underscore the potential of persistent homology and graph theoretical analysis of the whole-brain connectome in detecting alterations in structural connectivity patterns specific to different types of brain tumors.

Keywords: Diffusion-weighed MRI · Brain Connectome · Persistent Homology · Graph Theory · Classification

1 Introduction

Diffusion-weighted magnetic resonance imaging (DWI) tractography is a revolutionary imaging technique that allows for non-invasive reconstruction of the brain's WM fibre tracks at macro scale [2]. Tractography maps connections

Supplementary Information The online version contains supplementary material available at https://doi.org/10.1007/978-3-031-73967-5_4.

C. Chen et al. (Eds.): TGI3 2024, LNCS 15239, pp. 33–42, 2025.
https://doi.org/10.1007/978-3-031-73967-5_4

between different brain regions by tracing the diffusion of water molecules along axonal pathways, offering insights into the brain's structural connectivity and tissue micro-structure. This method has been significantly contributed across various neurological contexts, including aging, development, and disease characterization [8,18]. In this paper, we elucidate how tractography serves as a cornerstone for quantitative analysis in assessing changes of the brain structural connectivity for brain tumor classification problem. Recent advances in genome-based molecular investigations and clinical studies have revealed different sub-types of brain tumor. While numerous studies have explored various imaging features and their correlations with pathological and molecular characteristics, especially for tumor segmentation and classifying different grades of glioma, limited research has investigated the changes in structural connectivity of whole-brain due to tumor. Tumor often disrupt the normal architecture of brain by infiltrating healthy tissue, leading to changes in their connectivity patterns. Since, DWI tractography allows for visualization of WM tracts in the brain, in the context of studying brain tumor, tractography offers a unique opportunity to examine the changes in structural connectivity due to tumor with the surrounding brain tissues. The present study conducts a comprehensive topological data analysis and graph theoretical analysis on DWI whole-brain connectome matrix to differentiate between 3 study groups: (i) glioma, (ii) meningioma and (iii) healthy control. The main purpose is to distinguish these two types of brain tumors having distinct origins, characteristics, and treatment approaches. Meningiomas originate from meninges, the protective membranes surrounding brain and spinal cord, while gliomas arise from glial cells, which provide support and protection to neurons in brain. Differentiating between meningioma and glioma is important for diagnosis and treatment planning, as treatment strategies and prognoses differ between these two tumor sub-types. Meningiomas are typically benign, slow-growing which typically exhibit more localized growth and minimal invasion into surrounding tissue, thus, often amenable to surgical resection with favorable outcomes. In contrast, gliomas originates from glial cells are characterized by infiltrative growth and can vary widely in aggressiveness and prognosis, necessitating a more nuanced approach to treatment, including surgery, radiation therapy, and chemotherapy.

In the current study, we focus on applying a biphasic approach to differentiate meningiomas and gliomas using DWI brain connectome. Initially, we utilize persistent homology to capture the differences in topological characteristics of the brain connectome between the study groups. Persistent homology is crucial for identifying distinct topological features that might be overlooked by traditional graph measures, providing a deeper insight into the topological structure and its alterations due to different tumor types. This method enhances our ability to distinguish between the study groups by uncovering unique topological traits. Additionally, our study employs DWI connectome-based graph measures to classify the study groups. The importance of computing graphical features lies in their ability to capture the subtle local alterations in brain structural connectivity induced by various types of tumors. This dual approach offers a comprehensive analysis in examining the structural differences in brain connectomes among healthy individuals, and patients with meningioma and glioma, making this study a significant step forward in the research of connectome-based analysis of brain tumors. The block schematic of the proposed methodology is shown in Fig. 1.

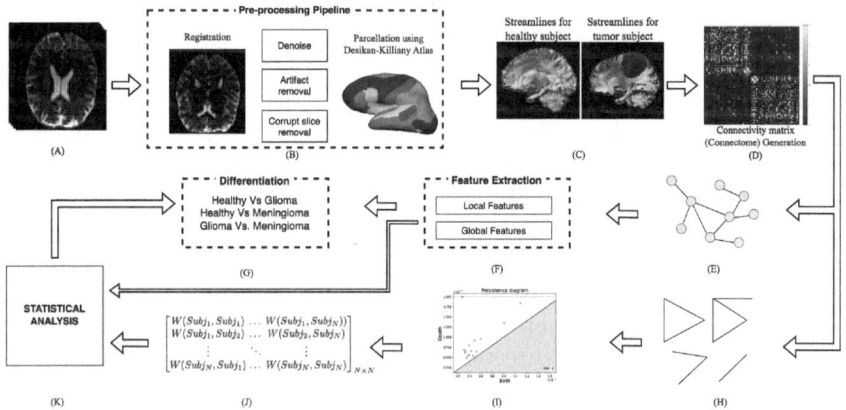

Fig. 1. Block schematic of the proposed study

2 Materials and Methods

2.1 Dataset and MRI Preprocessing

Publicly available Brain Tumor Connectomics (BTC) Dataset [13] is utilized that include pre-operative data of 25 patients who were diagnosed with (i) glioma ($n = 11$, mean age = 47.5y, SD = 11.3; M:F = 7:4), developing from glial cells, and (ii) meningioma ($n = 14$, mean age=60.4y, SD = 12.3; M:F = 3:11), developing in the meninges. A total 11 healthy control subjects (mean age = 58.6y, SD = 10.3; M:F = 7:4) are also incorporated for comparison purpose. From all participants, two types of MRI scans are utilized: (i) T1-W structural MRI (160 slices, TR = 1750 ms, TE = 4.18 ms, voxel size $1 \times 1 \times 1mm$), and (ii) a multishell high-angular resolution diffusion-weighted MRI (DWI) scan was acquired (60 slices; TR = 8700 ms; TE = 110 ms; 101 diffusion directions, voxel size = $2.5 \times 2.5 \times 2.5$mm, b-values = $0, 700, 1200, 2800\,\text{s/mm}^2$). In addition, two DWI $b = 0\,\text{s/mm}^2$ images are used with reversed phase-encoding blips for the purpose of correcting susceptibility-induced distortions. Further details regarding characteristics of the participants and MRI data acquisition can be found in paper by Hannelore Aerts et.al [13].

Structural MRI preprocessing utilized the automated FreeSurfer pipeline (version 7.1.0), which parcellated the cerebral cortex into gyral and sulcal structures based on 84 regions of interest (ROIs) from the Desikan-Killiany atlas (DK atlas) [9]. The labels of DK ROIs are listed in the github repository[1] The pre-processing steps are conducted using MRtrix toolbox [21] for DWI denoising, Gibbs' ringing artifacts removal. Additionally, FSL (version 6.0.7.7) libraries are also used for preprocessing. It includes brain extraction with the brain extraction tool, motion correction, correction for susceptibility-induced distortion using topup and correction for eddy-current distortion.

2.2   Whole-Brain Probabilistic Tractography

Anatomically-constrained probabilistic tractography [19] is conducted using the MRtrix3 software package [21]. Initially, five-tissue-type segmentation is performed, delineating cortical gray matter, subcortical gray matter, WM, cerebrospinal fluid, and other pathological tissues. Subsequently, constrained spherical deconvolution algorithm is used to estimate voxel-wise fibre orientation densities for each tissue type, providing insight into diffusion characteristics along multiple directions within a voxel. Since, MRtrix allows for the estimation of multiple crossing fibres within a single voxel, it can effectively disentangle the diffusion signal into multiple directions. Finally, whole-brain tractography is executed, initiating from the voxels at the gray matter-WM interface as seed points: the initial positions from which streamlines representing WM pathways are propagated throughout the brain. For each subject, a total of 10,000,000 streamlines are generated. The default values for the maximum length of streamlines and the angular threshold are set to 250mm and 0.06 radians, respectively, that determine when streamline propagation terminates. These parameters collectively influence several aspects of tractography, encompassing seeding, propagation, and termination criteria, thereby aiding in the delineation of WM pathways in the brain. Following the creation of the streamline map, a *connectome matrix* is generated, illustrating the number of streamlines connecting distinct brain regions (in this case, 84 ROIs of the DK atlas). However, computing DWI connectomes using MRtrix software is computationally very expensive, necessitating the use of high-performance computing (HPC) resources. Specifically, we employed an Intel(R) Xeon(R) Gold 6240 CPU @ 2.60 GHz with dual CPUs and 192 GB of memory for this purpose.

2.3   Persistent Homology on Whole-Brain Connectome

Persistent homology is a powerful topological data analysis approach that provides a robust framework for analyzing topological features of data, particularly in the context of shape and structure. At its core, persistent homology aims to capture the evolution of topological features across different spatial scales by constructing a sequence of topological spaces based on the input data. It reveals how specific topological characteristics evolve as we observe these spaces at different levels of detail. In the context of medical image analysis, persistent homology is used for analysis of endoscopy [10], analysis of brain networks for

---

[1] Link to code: https://github.com/blackpearl006/TGI3-2024/.

differentiating various types of brain disorders [14,20], analysis of visual brain networks [3], and detecting transition between states in EEG [15], identifying epileptic seizures [7]. In this study, we exploit the information encoded in *persistence diagram* to analyze DWI brain connectome of healthy individuals and patients having meningioma and glioma. Here, we provide a brief explanation of how persistence diagrams are constructed based on graph filtrations. Further details on persistence diagram can be found in literature [11]. The 0-th ordinary persistence diagram (H_0) captures the birth and death of connected components, known as 0-dimensional homological features. This is done using a sublevel set filtration that computes the birth and death of homology classes in a sequence of homology groups [11]. To capture the 1-dimensional features (loops), we compute the dimension-1 extended persistence diagram (H_1), which incorporates information from both sublevel set and superlevel set filtrations. The persistence diagram encodes the birth and death of 1-dimensional homology classes that persist throughout the sequence of absolute homology groups in the sublevel set filtration. These essential homology classes die in the relative homology groups of the superlevel set filtration and are represented as points in H_1 [3,11]. The quantification of persistence diagrams utilizes the *Wasserstein distance*, a metric that determines the dissimilarity (contributed by higher distance values) between two persistence diagrams. Specifically, it measures the extent of transformation needed for one persistence diagram to closely match another, providing a metric for comparing the topological structures of DWI connectome. In this study, Wasserstein distance is computed across all considered subjects for each of the Betti descriptors in dimension-0 (H_0) and dimension-1 (H_1). To compute the distance elements of two persistence diagrams X and Y one-to-one (bijection η) are matched. Mathematically, it is defined as,

$$W_{q,p}(X,Y) = \left[\inf_{\eta: X \to Y} \sum_{x \in X} ||x - \eta(x)||_\infty^q \right]^{\frac{1}{q}} \tag{1}$$

2.4 Graph-Based Feature Extraction

The DWI whole-brain connectome matrices are also used to compute features based on graph theory, with edges indicating the strengths of fiber connectivity between all pairs of 84 ROIs (nodes). The Brain Connectivity Toolbox from MATLAB is utilized for this purpose [17]. Here, we delve into graph-based characteristics that unveil both the functional integration and segregation aspects of DWI structural connectivity, with the goal of understanding the changes in the individual brain regions for different brain tumors. Global features of a graph provide insights into the functional segregation and functional integration of information flow within the brain network [12,17]. In the present study, 4 global features - *transitivity, modularity, characteristic path length, and density* - and 9 local features - *clustering coefficient (CC), nodal degree (Deg), betweenness centrality (BC), local efficiency (LE), eigenvector centrality (EVC), participation coefficient (PC), diversity coefficient (DC), gateway coefficient (GC), and strength (Str)*, are extracted. Detailed descriptions of all the considered 13 features can be found in paper [17].

3 Results and Discussion

This study conducts DWI brain connectome analysis of brain tumors with probabilistic tractography, generating WM streamlines that connect 84 gray matter regions defined by the DK atlas. The generation of streamlines for one healthy and one tumor subject are shown in Fig. 1(C). As seen from the figure, the absence of streamlines is evident within the tumor affected region. This observation is indicative of the disruption of WM tracts within the brain due to the presence of the tumor. Figure 1(D) displays the resulting connectome, illustrating the number of streamlines connecting 84 DK atlas regions in the brain. Increased structural connectivity within each hemisphere is noticeable and depicted in the diagonal boxes (bright yellow spots), indicating efficient communication between neighboring brain regions. Additionally, brighter spots in the off-diagonal boxes indicate increased structural connectivity between homologous regions of the hemispheres which is essential for various cognitive and motor functions.

3.1 Persistent Homology Based Analysis

The entire brain connectome is used to compute persistent diagrams, illustrated in Fig. 2. The persistence diagrams reveal clear differences in topological features among the groups: Control, Meningioma, and Glioma, for both dimension-0 and -1. In the control group, a high density of points near the origin in dimension-1 indicates that many loops are born and die quickly. In contrast, the tumor groups exhibit points that are more dispersed and away from diagonal line, indicating a wider range of feature persistence. This suggests that the tumor groups have more varied and potentially significant topological features that persist over a broader range of filtration values, likely due to the more complex and disrupted network structure-reflecting the underlying pathological changes in the brain network structure due to presence of tumor. As seen from Fig. 2, the dimension-1 persistence diagram displays a block-type structure due to several features being born and dying at similar filtration steps. Generally, this is seen to occur under specific conditions:

(i) Dense connectivity or full connected network: In a dense matrix or fully connected graph, many entries in the matrix may surpass a given threshold simultaneously. As a consequence, multiple features can be created or destroyed at the same filtration step, resulting either vertical or horizontal alignments in the persistence diagram. A proof-of-concept demonstrating this behavior is provided in the *Supplementary material, Fig. 1*.

(ii) Clusters in data: When non-zero values in the connectivity matrix are concentrated around certain levels, during the filtration process features tend to be created or destroyed at these clustered values. This results in multiple features being born or dying at the same filtration values, leading to aligned points

in the persistence diagram. A proof-of-concept demonstrating this behavior is shown kernel density estimate (KDE) and the corresponding histogram of brain connectome at the right panel in Fig. 2.

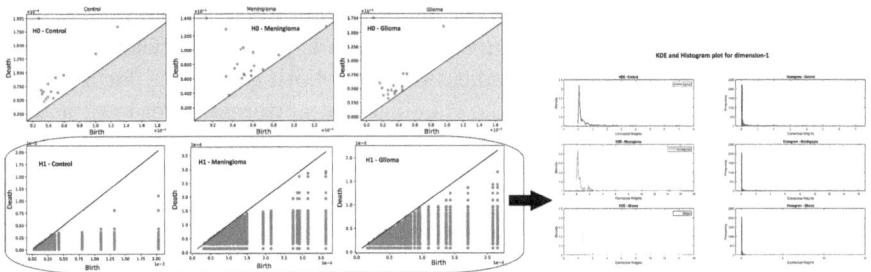

Fig. 2. The persistence diagram in dimension-0 (top row in left panel) and dimension-1 (bottom row in left panel) as obtained from whole brain connectome, for one representative subject of control (first column), glioma (second column) and meningioma (third column). The kernel density estimate (KDE) plot and the corresponding histogram for the dimension-1 persistence diagram is shown at the right panel.

Wasserstein distance as shown in Fig. 3 is used to quantify the dissimilarities in persistence diagrams between groups. Wilcoxon rank-sum test was conducted at 95% C.I. to determine if the differences in Wasserstein distance are statistically significant between study groups. For dimension-1, the Wasserstein distance shows statistically significant differences between HC vs. Meningioma ($p < 0.001$) and Glioma vs. Meningioma ($p < 0.001$), illustrated in the violin plot in Fig. 3. However, for dimension-0, no significant differences are observed between study groups.

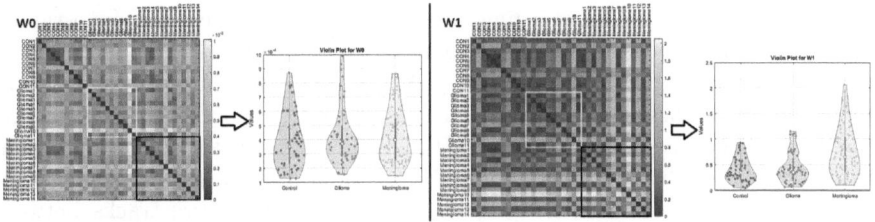

Fig. 3. Visualization of Wasserstein distance and the corresponding violin plot for dimension-0 (first and second column) and dimension-1 (third and fourth column), across all subjects to find the statistical significance($p < 0.05$) across study groups.

3.2   Graph Theory Based Analysis

A total of 13 graph-based features from the whole-brain connectome are used as inputs for various machine learning (ML) models to classify (i) Glioma Vs. HC, (ii) Meningioma Vs. HC, and (iii) Glioma Vs. Meningioma. Ensemble-based classifiers such as XGBoost, RUSBoost, and Random forest are employed. While using local features, recursive feature elimination is performed before classification. The performance of ML classification using both global and local features is presented in Table 1. However, the classification performance of local features are found to be superior than using global features. Among the three different classifiers, the highest accuracies are achieved using RUSBoost, with 88% for classifying HC Vs. meningioma. In the classification of tumor sub-types, an accuracy of 80% is achieved in distinguishing meningioma Vs. glioma using RUS-Boost and Random forest. Wilcoxon ransum test is conducted at 95% C.I. to assess the significant differences in local graph feature characteristics for each classification scenarios. This analysis aimed to identify specific brain ROIs where alterations in the graphical properties of WM connectivity are statistically significant (*Supplementary material Table 1*). Among 9 different local features, local efficiency and strength emerge as the most important features for distinguishing between the two classes. Visualization of significant ROIs for these two features is shown in *Supplementary material, Fig. 2*. However, no significant differences are observed in connectome-based global graph features between groups. While numerous studies have achieved comparable accuracy in brain tumor classification [1,4–6,16], none have utilized topological features for this task. Our findings are comparable to the insights provided by baseline study [13], highlight the effectiveness of topological data analysis in brain tumor classification. Unlike the baseline study, which focused on functional connectivity prediction and tumor region differentiation without classifying tumor types, our results provide a quantitative measure, demonstrating competitive performance in distinguishing different type of brain tumor by integrating DWI connectomes with persistent homology.

The current study adopts a unique and distinct approach that investigates the application of both DWI connectome-based graph measures and topological measures using persistent homology in order to differentiate brain tumors having different origins. The significance of this study lies in its ability to provide a comprehensive understanding of the brain's structural connectivity alterations associated with different types of brain tumor. By analyzing connectome data, which represents the connections strength in WM fibre tracks between brain regions, clinicians and researchers can gain insights into the underlying pathology of different types of brain tumor. Connectome-based graph-theoretical approaches offer the advantage of capturing both global and local alterations in brain connectivity, allowing for a more nuanced understanding of tumor-related changes across the entire brain. The incorporation of persistent homology adds a layer of depth to the study by capturing and quantifying subtle topological characteristics of the brain connectome. This method identifies topological features in the brain network that persist across different spatial scales, offering a

Table 1. Classification performance using graph-based global and local features

Feature	Classifier	Subjects	Precision	Recall	F1-score	Accuracy	AUC
Global	XGBoost	HC Vs Glioma	0.46	0.55	0.50	45%	0.45
		HC Vs Meningioma	0.68	0.93	0.79	72%	0.69
		Glioma Vs Meningioma	0.57	0.57	0.57	52%	0.51
	RUSBoost	HC Vs Glioma	0.55	0.55	0.55	56%	0.55
		HC Vs Meningioma	0.68	0.93	0.79	72%	0.69
		Glioma Vs Meningioma	0.64	0.50	0.56	64%	0.57
	Random Forest	HC Vs Glioma	0.42	0.45	0.43	50%	0.41
		HC Vs Meningioma	0.63	0.71	0.66	68%	0.58
		Glioma Vs Meningioma	0.57	0.57	0.57	60%	0.51
Local	XGBoost	HC Vs Glioma	0.78	0.64	0.70	**73%**	**0.79**
		HC Vs Meningioma	0.86	0.86	0.86	**84%**	**0.92**
		Glioma Vs Meningioma	0.73	0.79	0.76	**72%**	**0.90**
	RUSBoost	HC Vs Glioma	0.80	0.78	0.76	**77%**	**0.93**
		HC Vs Meningioma	0.92	0.86	0.89	**88%**	**0.94**
		Glioma Vs Meningioma	0.85	0.79	0.81	**80%**	**0.90**
	Random Forest	HC Vs Glioma	0.78	0.64	0.70	**77%**	**0.81**
		HC Vs Meningioma	0.81	0.93	0.87	**84%**	**0.91**
		Glioma Vs Meningioma	0.85	0.79	0.81	**80%**	**0.91**

more comprehensive view of how brain connectivity is altered in the presence of meningiomas versus gliomas. This analysis not only enhances the differentiation between tumor types but also contributes to our understanding of the complex network changes induced by these tumors. While further studies such as utilizing persistent homology features for tumor classification, longitudinal analysis are necessary for broader applicability and conclusive inferences, the holistic view offered by connectome analysis in this study has potential to enhance personalized medicine strategies, thereby improving patient outcomes and quality of life. The findings from this study could potentially translate into clinical practice by providing neurosurgeons and oncologists with more precise diagnostic tools. The ability to accurately classify tumors having different origin, based on their unique topological signatures may guide surgical planning, treatment selection, and monitoring of treatment response.

Disclosure of Interests. The authors have no competing interests.

References

1. Amin, J., Sharif, M., Haldorai, A., Yasmin, M., Nayak, R.S.: Brain tumor detection and classification using machine learning: a comprehensive survey. Complex Intell. Syst. **8**(4), 3161–3183 (2022)

2. Basser, P.J., Pierpaoli, C.: Microstructural and physiological features of tissues elucidated by quantitative-diffusion-tensor MRI. JMR **213**(2), 560–570 (2011)
3. Bhattacharya, D., Sinha, N., Chattopadhyay, A., et al.: Image complexity based FMRI-bold visual network categorization across visual datasets using topological descriptors and deep-hybrid learning. arXiv preprint arXiv:2311.08417 (2023)
4. Bhattacharya, D., Sinha, N., Saini, J.: Detection of chromosomal arms 1P/19Q codeletion in low graded glioma using probability distribution of MRI volume heterogeneity. In: TENCON 2019 - 2019 IEEE Region 10 Conference (TENCON), pp. 2695–2699 (2019). https://doi.org/10.1109/TENCON.2019.8929255
5. Bhattacharya, D., Sinha, N., Saini, J.: Determining chromosomal arms 1P/19Q co-deletion status in low graded glioma by cross correlation-periodogram pattern analysis. Sci. Rep. **11**(1), 23866 (2021)
6. Bhattacharya, D., Sinha, N., Saini, J.: Radial cumulative frequency distribution: a new imaging signature to detect chromosomal arms 1p/19q co-deletion status in glioma. In: Singh, S.K., Roy, P., Raman, B., Nagabhushan, P. (eds.) CVIP 2020. CCIS, vol. 1376, pp. 44–55. Springer, Singapore (2021). https://doi.org/10.1007/978-981-16-1086-8_5
7. Caputi, L., Pidnebesna, A., Hlinka, J.: Promises and pitfalls of topological data analysis for brain connectivity analysis. Neuroimage **238**, 118245 (2021)
8. Ciccarelli, O.: Diffusion-based tractography in neurological disorders: concepts, applications, and future developments. Lancet Neurol. **7**(8), 715–727 (2008)
9. Desikan, R., Killiany, R.: An automated labeling system for subdividing the human cerebral cortex on MRI scans into GYRAL based regions of interest. Neuroimage **31**, 961–980 (2006)
10. Dunaeva, O., Edelsbrunner, H.: The classification of endoscopy images with persistent homology. Pattern Recogn. Lett. **83**, 13–22 (2016)
11. Edelsbrunner, H., Harer, J.: Computational Topology - an Introduction. American Mathematical Society (2010)
12. Farahani, F.V., W.K., Lighthall, N.R.: Application of graph theory for identifying connectivity patterns in human brain networks: a systematic review. Front. Neurosci. **13**, 585 (2019)
13. Hannelore, A., Michael, S., Ben, J.: Modeling brain dynamics in brain tumor patients using the virtual brain. eNeuro **5**, 1–15 (2018)
14. Lee, H., Kang, H.: Persistent brain network homology from the perspective of dendrogram. IEEE Trans. Med. Imaging **31**(12), 2267–2277 (2012)
15. Merelli, E., Piangerelli, M.: A topological approach for multivariate time series characterization: the epileptic brain. In: In. Proc. BICT 2015, pp. 201–204 (2016)
16. Raja, R., Sinha, N.: Assessment of tissue heterogeneity using DTI and DKI for grading gliomas. Neuroradiology **58**, 1217–1231 (2016)
17. Rubinov, M., Sporns, O.: Complex network measures of brain connectivity: uses and interpretations. Neuroimage **52**, 1059–1069 (2010)
18. Shi, Y., Toga, A.W.: Connectome imaging for mapping human brain pathways. Mol. Psychiatry **22**(9), 1230–1240 (2017)
19. Smith, R., Tournier, J., Calamante, F., Connelly, A.: Anatomically-constrained tractography: improved diffusion MRI streamlines tractography through effective use of anatomical information. Neuroimage **62**, 1924–1938 (2012)
20. Stolz, B.J., Emerson, T.: Topological data analysis of task-based FMRI data from experiments on schizophrenia. J. Phys. Complex. **2**(3), 035006 (2021)
21. Tournier, J., Calamante, F., Connelly, A.: MRtrix: diffusion tractography in crossing fiber regions. Int. J. Imaging Syst. Technol. **22**, 53–66 (2012)

A Bispectral 3D U-Net for Rotation Robustness in Medical Segmentation

Arthur Chevalley[1,2]($\boxtimes$), Valentin Oreiller[2], Julien Fageot[2], John O. Prior[1], Vincent Andrearczyk[2], and Adrien Depeursinge[1,2]

[1] Nuclear Medicine and Molecular Imaging Department, Lausanne University Hospital (CHUV), Lausanne, Switzerland
`arthur.chevalley@hevs.ch`
[2] Institute of Informatics, HES-SO Valais-Wallis University of Applied Sciences and Arts Western Switzerland, Sierre, Switzerland

Abstract. Segmentation models achieved expert-level performance in a large variety of medical applications. However, their robustness to rotations is rarely discussed and can be crucial for clinical use with the risk of discarding subtle but diagnostically relevant anatomical structures. In medical images, complex structures appear in a wide range of positions and rotations, requiring rotation robustness. In this work, we investigate the robustness to rotations of a standard 3D nnU-Net in the context of two segmentation tasks: the hippocampus in MRI and the pulmonary airway system in CT. In addition, we introduce a 3D Locally Rotation Invariant (LRI) operator based on the bispectrum to achieve high robustness to input rotations. It is compared to a standard nnU-Net, a nnU-Net with extended rotational data augmentation and XEdgeConv, a state-of-the-art approach for RI. While all models performed similarly in terms of Dice score for right-angle rotations, the Bispectral U-Net outperformed other designs in the context of finer and more realistic rotations. Furthermore, the Bispectral U-Net and the XEdgeConv were more stable w.r.t. input rotation, i.e. the predictions are significantly more consistent across input rotations. Important inconsistencies of the nnU-Net were observed for lung airway segmentation, suggesting potential risks of using the model in clinical routine.

Keywords: Local Rotation Invariance · Robust 3D Segmentation · Convolutional Bispectral Network · Deep Learning · Medical Image Analysis

1 Introduction

Convolutional Neural Networks (CNN) are currently the workhorse for many medical image analysis tasks. These models must reach high performance and

Supplementary Information The online version contains supplementary material available at https://doi.org/10.1007/978-3-031-73967-5_5.

C. Chen et al. (Eds.): TGI3 2024, LNCS 15239, pp. 43–54, 2025.
https://doi.org/10.1007/978-3-031-73967-5_5

reliability for segmentation as errors can lead to severe clinical consequences. The nnU-Net [7] introduced a self-adapting framework reaching state-of-the-art performances. However, this framework relies solely on data augmentation to achieve rotation robustness, which may not be sufficient for accurately contouring biomedical structures appearing at a wide range of orientations, with global as well as local image rotations. Nevertheless, the impact of input rotations on the model's performance has been little studied to date. In this work, we first evaluate the performance robustness and rotation stability of the nnU-Net. We then propose a 3D extension of a 2D Rotation Invariant (RI) segmentation model based on the bispectral operator [12]. Note that the operator complexity, projection over the spherical harmonics rather than the circular ones, and the range of possible rotations make this step quite challenging. While rotations are only parameterized by one angle in 2D, 3D rotations are characterized by three angles, increasing their complexity and diversity by a cubic factor. We finally compare the performance and stability of these models with another RI model, XEdgeConv, on two segmentation tasks. Code will be made available.

2 Related Work

CNN's robustness to input rotations, a key property for various medical tasks, has seen limited investigation. Generally, rotational robustness is achieved with heavy data augmentation which does not pledge invariance. To address this, invariant or equivariant convolution layers were proposed. One strategy is to design group-equivariant convolution layers relying on rotated and reflected duplicates of all kernels to achieve equivariance [1,20]. Other developed specific parametric and/or steerable kernels [18] to create SE(3) equivariant networks using Circular [22] or Spherical Harmonics [4].

For texture classification, handcrafted Locally Rotation Invariant (LRI) operators achieved excellent results, e.g. local binary patterns [6] or steerable detectors [5,16]. Those operators have also been incorporated in 3D CNNs using SHs [2,11]. RI 3D point cloud segmentation was proposed [8,14,15,26] but only a few works focused on invariant segmentation of 3D images.

CubeNet [21] uses the group convolution in 3D to create rotation and translation equivariant CNNs using Klein's four- and tetrahedral-group, yielding a *Roto-translational group-convolution*. 3D-UCaps [10] proposed a 3D capsules pathway in addition to a standard CNN. However, the capsules do not fully encode rotation invariance as they reported similar segmentation performances robustness compared to a standard U-Net. XEdgeConv [17] uses recent advances in graph neural networks to construct a kernels from translation and permutation invariant graphs. Finally, [12] implemented a 2D LRI U-Net based on the bispectrum operator, which we extend to 3D image segmentation in this work.

3 Methods

This section presents the mathematical background for the proposed Bispectral U-Net for 3D images. By embedding the bispectrum operator in a convolutional layer, it is convenient to compare the effect of various convolution strategies for a given network architecture, e.g. a *standard* 3D convolution for the nnU-Net model or another RI method such as XEdgeConv [17]. Compared to the 2D implementation [12], the projection is made on the spherical harmonics to be applied to 3D images, thus increasing the computational complexity.

3.1 Notations

The 3D images considered are defined as functions $I(\boldsymbol{x}) \in L_2(\mathbb{R}^3)$ where $I(\boldsymbol{x})$ is the image intensity at the location $\boldsymbol{x} = (x_1, x_2, x_3) \in \mathbb{R}^3$. The spherical coordinates are defined as (ρ, θ, ϕ) where $\rho \geq 0$ is the radius, $\theta \in [0, \pi]$ the elevation angle and $\phi \in [0, 2\pi[$ the horizontal plane angle. In addition, on $\mathbb{R}^3$, the unit sphere is defined as $\mathbb{S}^2 = \{\boldsymbol{x} \in \mathbb{R}^3 : ||\boldsymbol{x}||_2 = 1\}$. Functions on the sphere are given as $f \in L_2(\mathbb{S}^2)$ using the spherical coordinates. The inner-product for $f, g \in L_2(\mathbb{S}^2)$ is defined by $\langle f, g \rangle_{L_2(\mathbb{S}^2)} = \int_0^\pi \int_0^{2\pi} f(\theta, \phi) g(\theta, \phi) \sin(\theta) \mathrm{d}\phi \mathrm{d}\theta$. The triangle function is defined as $\mathrm{tri}(\boldsymbol{x}) = 1 - |\boldsymbol{x}|$ if $|\boldsymbol{x}| < 1$ and $\mathrm{tri}(\boldsymbol{x}) = 0$ otherwise. Finally, the Kronecker product is denoted by $\otimes$ and the Hermitian transpose by †.

3.2 Bispectrum Operators

In addition to the global rotation and translation equivariance properties provided by any LRI operator [2], the bispectrum operator is sensitive to directional patterns and complete, i.e. functions with identical bispectrum are rotated versions of each other [9]. As this work focuses on analysing rotation robustness, we only recall essential definitions as extended details are provided in [11].

The bispectrum is computed using the Fourier transform on the sphere relying on Spherical Harmonics (SH). The family of SHs, made of functions Y_n^m for a degree $n \in \mathbb{N}$ and order m with $-n \leq m \leq n$, is known to form an orthonormal basis of $L_2(\mathbb{S}^2)$. Therefore, any function $f \in L_2(\mathbb{S}^2)$ can be projected onto the SH basis following the inner product

$$F_n^m = \langle f, Y_n^m \rangle_{L_2(\mathbb{S}^2)}, \tag{1}$$

in which case $f = \sum_{n \geq 0} \sum_{-n \leq m \leq n} F_n^m Y_n^m$. The Spherical Fourier (SF) vector for a given degree n is grouping the coefficients of all orders m as

$$\mathcal{F}_n = [F_n^{-n}..F_n^0..F_n^n]. \tag{2}$$

Finally, following [19], a bispectrum coefficient of degree l, $|n - n'| \leq l \leq n + n'$, of any function $f \in L_2(\mathbb{S}^2)$ and $n, n' \geq 0$ can be obtained with the operator $\mathcal{B}$ and is defined as

$$b_{n,n'}^l(f) = [\mathcal{F}_n \otimes \mathcal{F}_{n'}]C_{nn'}\tilde{\mathcal{F}}_l^\dagger = \mathcal{B}\{\mathcal{F}_n, \mathcal{F}_{n'}, \mathcal{F}_l\}, \tag{3}$$

where $\mathcal{F}_n \otimes \mathcal{F}_{n'}$ is a $1 \times (2n+1)(2n'+1)$ vector and $C_{nn'}$ is the $(2n+1)(2n'+1) \times (2n + 1)(2n' + 1)$ Clebsh-Gordan matrix containing the namesake coefficients. $\tilde{\mathcal{F}}_l$ contains the SF vector of degree l. It is zero-padded to match $C_{nn'}$ size and allows to *select* only the rows associated with the lth degree.

Bispectrum Operators for 3D Images. While Eq. (2) applies to functions defined on the sphere, this work is interested in its application to 3D images. This requires extending SH bases to 3D volumes. The so-called Solid SHs are created by multiplying SHs with compactly supported radial profiles for each degree $h_n(\rho)$ [11]. Solid SHs of degree n and order m evaluated on the Cartesian grid are defined as

$$\kappa_n^m(\boldsymbol{x}) = \kappa_n^m(\rho, \theta, \phi) = h_n(\rho)Y_n^m(\theta, \phi). \tag{4}$$

From this equation, SF maps can be created for each degree by convolving the image with the solid SHs as

$$\mathcal{F}_n(\boldsymbol{x}) = [(I * \kappa_n^m)(\boldsymbol{x})]_{m=-n}^{m=n}. \tag{5}$$

Note the slight abuse of notation as $\mathcal{F}_n(\boldsymbol{x})$ is the local projection or the image around a position $\boldsymbol{x}$ to a function defined on the sphere before being projected onto the SHs basis [11].

Finally, for any $I \in L_2(\mathbb{R}^3)$ and $\boldsymbol{x} \in \mathbb{R}^3$, the bispectrum image operator of degree $n, n' \geq 0$ and $|n - n'| \leq l \leq n + n'$ can be created from Eq. (3) and (5) as

$$\mathcal{G}_{n,n',l}\{I\}(\boldsymbol{x}) = \mathcal{B}\{\mathcal{F}_n(\boldsymbol{x}), \mathcal{F}_{n'}(\boldsymbol{x}), \mathcal{F}_l(\boldsymbol{x})\}. \tag{6}$$

This operator inherits the invariance properties (e.g. LRI) of the bispectrum [11].

3.3   Bispectral LRI Layer Implementation

The bispectrum operator implementation requires setting the maximal SH decomposition degree $N \geq 0$. The computed coefficients are restrained to limit the computational cost, i.e. complexity of $\mathcal{O}(N^3)$. In addition, N is limited by the discretization of the solid SH on a cubic kernel, similarly to a Nyquist frequency, by $N \leq \frac{\pi c}{4}$ where c is the kernel size [2]. As stated in [11], only the components satisfying $0 \leq n \leq n'$ and $0 \leq n + n' \leq N$ are kept as $b_{n,n'}^l(f)$ and $b_{n',n}^l(f)$ are proportional independently of f. In addition, only the indices where the sum $n + n' + l$ is even are kept as the coefficients are observed to be zero otherwise.

The implementation of the solid SHs is done in multiple steps. Firstly, the radial profiles $h_n(\rho)$ are constructed as linear combinations of radial functions $\psi_j(\rho)$, Eq. (7) LHS. For this work, the radial functions were set to $\psi_j(\rho) = \text{tri}(\rho - j)$. The radial profiles are then evaluated on a Cartesian grid for discretization, as in [2], and normalised. Finally, they are multiplied with the SHs to create the kernel Eq. (7) RHS as

$$h_n^{i,o}(\rho) = \sum_{j=0}^{J} w_{n,j}^{i,o}\psi_j(\rho) \xrightarrow{Eq.\ (4)} \kappa_{n,m}^{i,o} = \left(\sum_{j=0}^{J} w_{n,j}^{i,o}\psi_j(\rho)\right)Y_n^m(\theta, \phi), \tag{7}$$

where $w_{n,j}^{i,o}$ are the trainable parameters of the model. J corresponds to the number of radial profiles, i.e. half the kernel size. The indices i and o iterate over $[1, ..., C_{in}]$ and $[1, ..., C_{out}]$ representing the number of input and output channels of the layer. Each bispectral convolution is performed in four steps. First, the feature maps, i.e. SF maps are computed as a convolution

$$\mathcal{F}_n^o(\boldsymbol{x}) = \sum_{i=1}^{C_{in}} [(y_i * \kappa_{n,m}^{i,o})(\boldsymbol{x})]_{m=-n}^{m=n}, \tag{8}$$

with y_i the ith channel of the previous feature maps and $\kappa_n^{i,o}$ the kernel described in Eq. (7). The indices i and o iterate over all input and output channels. The second step is to compute Eq. (6), using the SF maps of each degree $\{\mathcal{F}_n^o(\boldsymbol{x}), \mathcal{F}_{n'}^o(\boldsymbol{x}), \mathcal{F}_l^o(\boldsymbol{x})\}$. We consider the LRI output feature maps computed at each layer via the multichannel bispectrum operator

$$\mathcal{G}_{n,n',l}^o\{\boldsymbol{y}\}(\boldsymbol{x}) = [\mathcal{F}_n^o(\boldsymbol{x}) \otimes \mathcal{F}_{n'}^o(\boldsymbol{x})]C_{nn'}\tilde{\mathcal{F}_l^o}(\boldsymbol{x})^\dagger, \tag{9}$$

where $\boldsymbol{y}(\boldsymbol{x}) = [y_1(\boldsymbol{x}), \ldots, y_{C_{in}}(\boldsymbol{x})]$ and $\mathcal{F}_n^o(\boldsymbol{x})$ is given by Eq. (8). Note that strictly speaking, the operator Eq. (9) is not a bispectrum operator in the sense of Eq. (6) due to the sum in Eq. (8). It, however, inherits the bispectrum operator's equivariance properties, which we demonstrate in the supplementary material.

A non-linearity of the form $\sigma(x) = \text{sign}(x)\log(1 + x)$ is applied to avoid vanishing and exploding gradients due to the sizes of the bispectral feature maps. A bias is also added before applying a ReLU function. Finally, the features maps are convolved with a standard $1 \times 1 \times 1$ convolution to reduce the number of output channels to the desired C_{out}.

3.4   Datasets

The evaluation was conducted on two datasets pre-processed using nnU-Net [7] pipeline. The first one is the HippoCampus (HC) segmentation task of the *Medical Segmentation Decathlon* dataset[1] [3]. This dataset consists of 260 3D Magnetic Resonance Imaging (MRI) volumes with two classes contoured, i.e. the HC head as well as the union of its body and tail. 80% of the data (208 images) are used for training and the remaining 20% (52 images) for testing. The training set is split in four folds with 75% for training (156 images) and the remaining 52 images for validation. A second dataset, the Airway Tree Modeling 2022 (ATM22)[2] [13,23–25,27] dataset was also selected as it may be more prone to rotation sensibility. Its main task is to segment the pulmonary airway tree on Computed Tomography (CT) images. The first pre-processing step is to resample images to have an isotropic sampling, discarding images with a spacing difference larger than 0.5, resulting in 288 usable images. 75% (215 images) are used for training while the remaining 25% (73 images) are selected for testing. A subset of 22 test images were randomly selected to limit computational load when applying test time image rotations required to evaluate robustness (see Sect. 3.6). The training set is split again in two folds with 156 training and 59 validation images. For both datasets, the test results are the averaged model's output of all folds producing a more robust estimate of the model performances.

---

[1] http://medicaldecathlon.com/, March 2024.
[2] https://atm22.grand-challenge.org/, March 2024.

3.5   Network Details

The network architecture was automatically generated by the nnU-Net framework [7]. The encoder path was composed of four modules each containing two convolutional layers with a $3 \times 3 \times 3$ kernel. Every layer was constituted of a convolution followed by batch normalisation and LeakyRELU, with a negative slope of 0.01. Depending on the model tested, i.e. either the standard convolution, XEdgeConv or the proposed bispectral, the type of convolution was selected accordingly in each of those layers. Between modules, a max pooling layer, with a kernel size of $2 \times 2 \times 2$ and a similar stride, was used to reduce the dimensionality. The decoder path comprised three modules preceded by a trilinear upsampling. Finally, the prediction was computed with a $1 \times 1 \times 1$ standard convolution and a softmax activation, then binarized with argmax. For HC, a patch size of $40 \times 40 \times 40$ was used based on nnU-Net implementation [7] whereas for ATM22, a patch size of $56 \times 56 \times 56$ was selected based on GPUs' memory limitation. Even though this patch size is rather low to segment the whole airway tree, our main interest is to investigate the rotation stability and not the absolute segmentation performance. Compared to other models, the number of base features of the Bispectral U-Net, i.e. the input's number of channels, is reduced to eight to fit to the available memory. The network was trained using Dice and cross-entropy losses on an NVIDIA V100 for HC[3] and an NVIDIA A100 for ATM22[4]. The Bispectral U-Net used Adam optimization with a decaying learning rate starting at 1e-3. The maximum number of training epochs was set to 100 for the HC and 50 for ATM22 as the models reached a plateau. PyTorch 2.1.1 was used for all models.

Two other models were selected for comparison. nnU-Net [7] as a baseline and XEdgeConv [17], a state-of-the-art model regarding rotation stability in medical images. Both methods were trained with their default parameters, stochastic gradient descent with a learning rate decay starting at 1e−2. The standard nnU-Net data augmentation was used during training, including rotation randomly sampled between ±30° for each axis, for the three models: nnU-Net, XEdgeConv and Bispectral U-Net. However, as the nnU-Net's robustness to rotation solely relies on augmentation, a fourth model, referred to as nnU-Net Extended, was trained with an extended rotational augmentation range of ±180° for each axis.

3.6   Metrics and Evaluation

The rotational robustness and stability of each model are assessed by feeding multiple rotations of every test image and comparing the results between rotations for two tests. The first, referred to as *performance robustness* test, shows the actual segmentation performance of the model via the standard Dice similarity coefficient (DSC) where the rotated images are compared with the rotated ground truth. The second, referred to as the *rotational stability* test, directly evaluates the model's stability by comparing the prediction of each rotated image

---

[3] An epoch was computed in ≈ 15 min using up to 27 Gb for a batch size of two.
[4] An epoch was computed in ≈ 45 min using up to 70 Gb for a batch size of one.

with the non-rotated prediction. First, the network is fed with multiple rotations of the same image. Then, the inverse rotation is applied to each output probabilities map before being compared to the non-rotated probability map. Finally, the Root Mean Square Error (RMSE) between them measures consistency across rotations, e.g. a perfectly equivariant network will have an RMSE of zero.

As the volumes can be rotated in the whole space, a set of rotations must be selected as testing every angle is not feasible, i.e. graduation of ten degrees along each axis leads to more than tens of thousands of rotations compared to thirty-six in 2D. The first set is made of the 24 right-angle rotations according to Euler intrinsic angles $z - x' - z''$. However, since patient orientation is often controlled in medical images, i.e. a brain MRI or lung CT are unlikely to be upside down, right-angle rotations could be irrelevant since simple pre-processing steps could align all images. A set of realistic rotations is created by uniformly sampling 13 spherical coordinates within a cone of $45°$ angle, i.e. $\phi \in [-45°, 45°]$ and $\theta \in [0, 360°]$, referred to as $Cone^5$. Only rotations along z and x' are tested as z'' rotation induces limited variations for the right-angle test.

4   Results

Right-angle performance robustness and stability distribution are presented in Fig. 1a) and c) for both datasets. The nnU-Net performances, without training right-angle augmentation, are not shown in the figure as it is an unfair comparison given that the model has never seen such rotations during training.

Similarly, the $Cone$ rotations distributions are shown in Fig. 1b) and d). The classic nnU-Net is also included as the training range of rotations is including testing rotations.

The means of each distribution are reported in Table 1. For extreme rotations on the HC, all models except Bispectral U-Net and XEdgeConv second class ($p = 0.06$), are performing significantly different from each other in terms of performance robustness ($p \leq 0.015$). For rotational stability, Bispectral U-Net achieved significantly lower RMSE from all models ($p \leq 0.016$), while nnU-Net Extended and XEdgeConv are not significantly different ($p = 0.10$) for the second class.

For realistic rotations (i.e. $Cone$), all performance distributions except XEdgeConv and nnU-Net Extended (Class 1 $p = 0.67$, class 2 $p = 0.49$) are significantly different ($p = 0.02$). Regarding stability for realistic rotations, only the first class distributions of Bispectral U-Net and nnU-Net Extended ($p = 0.25$) are not significantly different. For extreme rotations on ATM22, nnU-Net Extended and XEdgeConv show no significant difference in performance ($p = 0.058$), but the Bispectral U-Net differs significantly from both (nnU-Net Extended $p = 0.0005$ and XEdgeConv $p = 0.0055$). All RMSE distributions have significantly different null p-values. For moderate rotations, only nnU-Net and nnU-Net Extended performance distributions are not different ($p = 0.39$). For

---

[5] A spline interpolation of order three was used when executing those rotations.

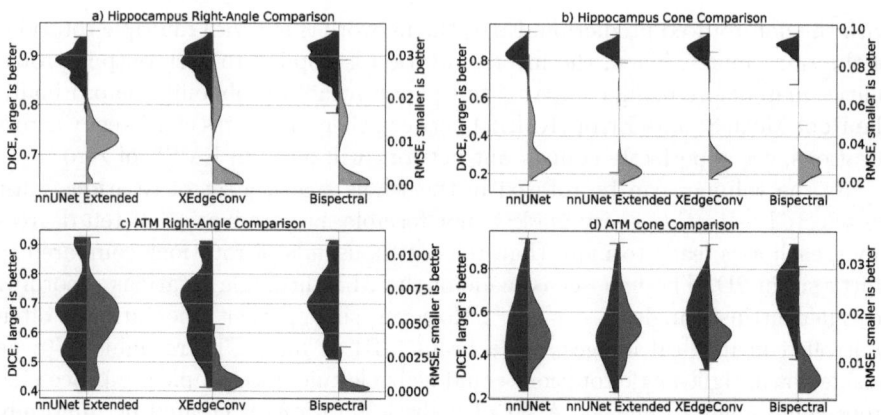

Fig. 1. DICE performance robustness of each model is shown on the left, darker, side of each violin plot while the right, lighter, side shows the rotation stability via RMSE. The violin plots are created using the metrics score for every image and every rotation. Sub-figures a), b) are the HippoCampus (HC) dataset with both classes aggregated in a single distribution. Sub-fig. c), d) shows Airway Tree Modeling (ATM22) distributions.

rotational stability, only XEdgeConv and nnU-Net Extended are not significantly different ($p = 0.82$).

Table 1. HC and ATM22 performance and stability metric means for the experiments with either 24 right-angle or 13 *Cone* rotations. Best results are highlighted in bold. Note that nnUNet results are not included in the table as they are worse than its extended version and compromise the table's readability. For the HC, both class's performances are regrouped in a single distribution.

Model	performance robustness (DSC)				rotational stability (RMSE)			
	Right-Angles		Cone		Right-Angles		Cone	
	HC	ATM22	HC	ATM22	HC	ATM22	HC	ATM22
Ext. nnU-Net	87.88%	64.93%	85.17%	50.97%	10.23e−3	4.96e−3	**2.64e−4**	1.74e−2
XEdge	88.38%	**67.42%**	85.21%	58.95%	8.13e−3	1.18e−3	3.09e−4	1.61e−2
Ours	**88.93%**	64.40%	**87.21%**	**68.89%**	**6.74e−3**	**0.75e−3**	2.72e−4	**0.85e−2**

5   Discussions and Conclusions

In this work, we investigated segmentation network stability w.r.t rotations. This question has seen limited investigations even though medical images contain complex structures in a wide range of rotations. We first investigated the stability nnU-Net to right-angle rotations. Note that the nnU-Net was trained with a wider range of rotations than the default (±180° instead of the basic ±30°). We then evaluated the same rotations on two RI networks trained without the extended rotation augmentation. One state-of-the-art approach, XEdgeConv, and the proposed 3D Bispectral U-Net based on its 2D alternative [12].

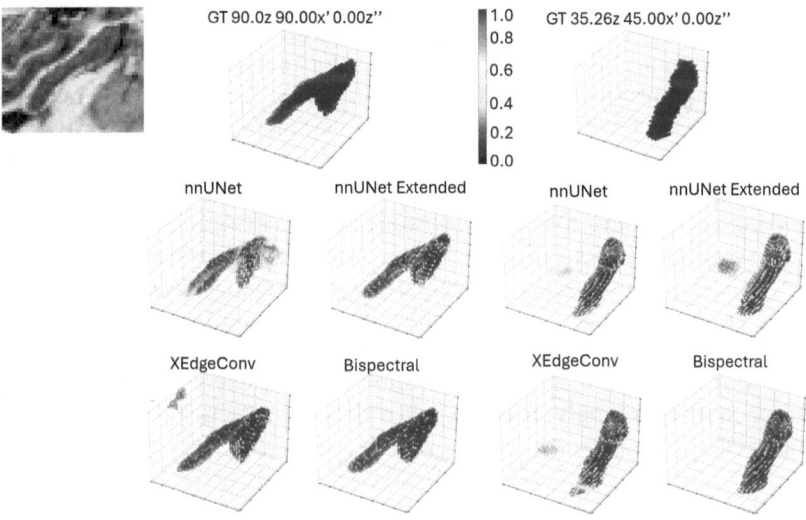

Fig. 2. Prediction confidence of each model for a right-angle rotation, left side, and a cone rotation, right side, of the HC dataset. Only hippocampus prediction is kept and background prediction is not shown. A slice of the input volume is shown in the top left corner.

When comparing the three networks for extreme rotations, clear benefits from the invariant networks can be observed w.r.t rotation stability while preserving the segmentation performance. Both invariant networks share very close performances with slightly better stability for the Bispectral U-Net. However, for smaller and more realistic rotations (i.e. *Cone*), the standard and extended nnU-Nets showed moderate performance robustness and rotational stability, with potentially serious clinical consequences. In this context, the Bispectral U-Net achieved significantly higher performance robustness and rotational stability for both datasets, which was even more marked for the ATM22 dataset.

When looking at the prediction confidence of different models in Fig. 2, both nnUNet models have much more voxels with lower confidence, i.e. greener, compared to other models. In addition, in some extreme cases, the models even detect the positive class outside of the main structure. When looking at right-angle rotations, the confidence of each model is higher, except the normal nnUNet.

The distinct observed trends between extreme and *Cone* rotations could be due to (i) more complex directional structures in ATM22 being more sensitive to smaller rotations and (ii) the induced effect of the rotations on the images is less diverse with right-angle rotations as they contain trivial symmetries. The difference with XEdgeConv can be explained as they use a maximum pooling across neighbouring voxels, which could lose information for smaller rotations. Our tests also show that the nnU-Net Extended is very competitive with RI networks regarding rotation stability for extreme rotations.

Our network shows several limitations. The bispectrum coefficients require large memory to be computed during training, as well as testing, thus limiting maximum decomposition degree, kernel size and patch size. However, such a bispectral model could be more than beneficial for datasets with smaller volumes and further computational optimisation. In addition, the training is much longer than other approaches even if the number of parameters is significantly lower as the bispectrum generates large matrices. Similarly, during inference, the prediction also requires an important memory and a rather long time to compute all the bispectrum coefficients. Nevertheless, we could successfully apply it to two real world clinical applications. Finally, when subject to right-angle rotations, the Bispectral U-Net did not show a generalised performance increase compared to the state-of-the-art model. Further improvements would be to conduct a hyperparameters search and test the network with larger kernel sizes and maximal degrees to extract more information. Similarly, the effect of smaller rotations should be more thoroughly investigated to know when a Bispectral U-Net would be preferred.

Acknowledgments. This work was partially funded by the Swiss National Science Foundation (SNSF) with the projects 205320_219430 and 205320_179069, the Swiss Cancer Research foundation with the project TARGET (KFS-5549-02-2022-R), and the Hasler Foundation with the project MSxplain number 21042.

Disclosure of Interests. The authors have no competing interests to declare that are relevant to the content of this article.

References

1. Andrearczyk, V., Depeursinge, A.: Rotational 3D texture classification using group equivariant CNNs. arXiv preprint arXiv:1810.06889 (2018)
2. Andrearczyk, V., Fageot, J., Oreiller, V., Montet, X., Depeursinge, A.: Local rotation invariance in 3D CNNs. Med. Image Anal. **65**, 101756 (2020)
3. Antonelli, M., et al.: The medical segmentation decathlon. Nat. Commun. **13**(1), 4128 (2022)
4. Esteves, C., Allen-Blanchette, C., Makadia, A., Daniilidis, K.: Learning SO(3) equivariant representations with spherical CNNs. In: Ferrari, V., Hebert, M., Sminchisescu, C., Weiss, Y. (eds.) ECCV 2018. LNCS, vol. 11217, pp. 54–70. Springer, Cham (2018). https://doi.org/10.1007/978-3-030-01261-8_4
5. Fageot, J., Uhlmann, V., Püspöki, Z., Beck, B., Unser, M., Depeursinge, A.: Principled design and implementation of steerable detectors. IEEE Trans. Image Process. **30**, 4465–4478 (2021)
6. Hadid, A.: The local binary pattern approach and its applications to face analysis. In: 2008 First Workshops on Image Processing Theory, Tools and Applications, pp. 1–9. IEEE (2008)

7. Isensee, F., Jaeger, P.F., Kohl, S.A., Petersen, J., Maier-Hein, K.H.: nnU-Net: a self-configuring method for deep learning-based biomedical image segmentation. Nat. Methods **18**(2), 203–211 (2021)

8. Jiang, M., Wu, Y., Zhao, T., Zhao, Z., Lu, C.: PointSIFT: a SIFT-like network module for 3D point cloud semantic segmentation. arXiv preprint arXiv:1807.00652 (2018)

9. Kakarala, R.: Completeness of bispectrum on compact groups. arXiv preprint arXiv:0902.0196 **1** (2009)

10. Nguyen, T., Hua, B.-S., Le, N.: 3D-UCaps: 3D capsules Unet for volumetric image segmentation. In: de Bruijne, M., et al. (eds.) MICCAI 2021, Part I. LNCS, vol. 12901, pp. 548–558. Springer, Cham (2021). https://doi.org/10.1007/978-3-030-87193-2_52

11. Oreiller, V., Andrearczyk, V., Fageot, J., Prior, J.O., Depeursinge, A.: 3D solid spherical bispectrum CNNs for biomedical texture analysis. arXiv preprint arXiv:2004.13371 (2020)

12. Oreiller, V., Fageot, J., Andrearczyk, V., Prior, J.O., Depeursinge, A.: Robust multi-organ nucleus segmentation using a locally rotation invariant bispectral U-Net. In: International Conference on Medical Imaging with Deep Learning, pp. 929–943. PMLR (2022)

13. Qin, Y., et al.: AirwayNet: a voxel-connectivity aware approach for accurate airway segmentation using convolutional neural Networks. In: Shen, D., et al. (eds.) MICCAI 2019. LNCS, vol. 11769, pp. 212–220. Springer, Cham (2019). https://doi.org/10.1007/978-3-030-32226-7_24

14. Rao, Y., Lu, J., Zhou, J.: Spherical fractal convolutional neural networks for point cloud recognition. In: Proceedings of the IEEE/CVF Conference on Computer Vision and Pattern Recognition, pp. 452–460 (2019)

15. Sun, X., Lian, Z., Xiao, J.: SRINet: learning strictly rotation-invariant representations for point cloud classification and segmentation. In: Proceedings of the 27th ACM International Conference on Multimedia, pp. 980–988 (2019)

16. Unser, M., Chenouard, N.: A unifying parametric framework for 2D steerable wavelet transforms. SIAM J. Imag. Sci. **6**(1), 102–135 (2013)

17. Weihsbach, C., Hansen, L., Heinrich, M.: XEdgeConv: leveraging graph convolutions for efficient, permutation- and rotation-invariant dense 3D medical image segmentation. In: Geometric Deep Learning in Medical Image Analysis, pp. 61–71. PMLR (2022)

18. Weiler, M., Geiger, M., Welling, M., Boomsma, W., Cohen, T.S.: 3D Steerable CNNs: learning rotationally equivariant features in volumetric data. Adv. Neural Inf. Process. Syst. **31** (2018)

19. Weiler, M., Hamprecht, F.A., Storath, M.: Learning steerable filters for rotation equivariant CNNs. In: Proceedings of the IEEE Conference on Computer Vision and Pattern Recognition, pp. 849–858 (2018)

20. Winkels, M., Cohen, T.S.: Pulmonary nodule detection in CT scans with equivariant CNNs. Med. Image Anal. **55**, 15–26 (2019)

21. Worrall, D., Brostow, G.: CubeNet: equivariance to 3D rotation and translation. In: Ferrari, V., Hebert, M., Sminchisescu, C., Weiss, Y. (eds.) ECCV 2018. LNCS, vol. 11209, pp. 585–602. Springer, Cham (2018). https://doi.org/10.1007/978-3-030-01228-1_35

22. Worrall, D.E., Garbin, S.J., Turmukhambetov, D., Brostow, G.J.: Harmonic networks: deep translation and rotation equivariance. In: Proceedings of the IEEE Conference on Computer Vision and Pattern Recognition, pp. 5028–5037 (2017)

23. Yu, W., Zheng, H., Zhang, M., Zhang, H., Sun, J., Yang, J.: BREAK: bronchi reconstruction by geodesic transformation and skeleton embedding. In: 2022 IEEE 19th International Symposium on Biomedical Imaging (ISBI), pp. 1–5. IEEE (2022)
24. Zhang, M., et al.: Multi-site, multi-domain airway tree modeling. Med. Image Anal. **90**, 102957 (2023)
25. Zhang, M., Zhang, H., Yang, G.Z., Gu, Y.: CFDA: collaborative feature disentanglement and augmentation for pulmonary airway tree modeling of COVID-19 CTs. In: Wang, L., Dou, Q., Fletcher, P.T., Speidel, S., Li, S. (eds.) MICCAI 2022. LNCS, vol. 13431, pp. 506–516. Springer, Cham (2022). https://doi.org/10.1007/978-3-031-16431-6_48
26. Zhang, Z., Hua, B.S., Rosen, D.W., Yeung, S.K.: Rotation invariant convolutions for 3D point clouds deep learning. In: 2019 International Conference on 3D Vision (3DV), pp. 204–213. IEEE (2019)
27. Zheng, H., et al.: Alleviating class-wise gradient imbalance for pulmonary airway segmentation. IEEE Trans. Med. Imaging **40**(9), 2452–2462 (2021)

Restoring Connectivity in Vascular Segmentations Using a Learned Post-processing Model

Sophie Carneiro-Esteves[1,2] (ID), Antoine Vacavant[1] (ID),
and Odyssée Merveille[2]([✉]) (ID)

[1] Université Clermont Auvergne, CNRS, SIGMA Clermont, Institut Pascal,
63000 Clermont-Ferrand, France
`antoine.vacavant@uca.fr`
[2] INSA-Lyon, Universite Claude Bernard Lyon 1, CNRS, Inserm,
CREATIS UMR 5220, U1294, 69100 Lyon, France
{`sophie.carneiro,odyssee.merveille`}`@creatis.insa-lyon.fr`

Abstract. Accurate segmentation of vascular networks is essential for computer-aided tools designed to address cardiovascular diseases. Despite more than thirty years of research, it remains a challenge to obtain vascular segmentation results that preserve the connectivity of the underlying vascular network. Yet connectivity is one of the key features of these tools. In this work, we propose a post-processing algorithm aiming to reconnect vascular structures that have been disconnected by a segmentation algorithm. Connectivity being a complex property to model explicitly, we propose to learn this geometric feature either through synthetic data or annotations of the application of interest. The resulting post-processing model can be used on the output of any supervised or unsupervised vascular segmentation algorithm. We show that this post-processing effectively restores the connectivity of vascular networks both in 2D and 3D images, leading to improved overall segmentation results.

Keywords: Blood vessels · segmentation · connectivity · deep learning · post-processing

1 Introduction

Blood vessel segmentation is a crucial step for various tasks such as blood flow simulation or 3D modeling, and enhancing our understanding of vascular networks physiology, and pathologies. However, segmenting blood vessels is challenging due to their thin and tortuous nature, making them easily altered by noise and artifacts. This often results in fragmented blood vessel segmentations which is a major problem for most downstream tasks.

Supplementary Information The online version contains supplementary material available at https://doi.org/10.1007/978-3-031-73967-5_6.

For over thirty years, methods have been introduced to enhance both the quality and the connectivity of blood vessel segmentation. Several unsupervised filtering approaches were first proposed. Vesselness filters [15] aims at enhancing the signal from blood vessels and decreasing the one from other non-tubular structures. These filters are designed to detect blood vessels at different scales, employing either a Gaussian-scale paradigm and Hessian-based features extraction [8,29], or a mathematical morphology approach using paths as structuring elements [19]. These filters are usually the first step of more complex segmentation pipelines [5,20,21].

However, determining hyperparameters for these filters can be challenging and there is no guarantee on the connectivity of the vascular tree. Alternative unsupervised methods, such as tracking [3] or minimal path methods [17] can ensure the structure connectivity. Nevertheless, these methods require a time-consuming user interaction to define seed points. All these approaches are further limited by having to explicitly model blood vessels.

Supervised methods, and in particular deep learning-based ones, offer the power to represent complex phenomena by learning implicit functions, provided there are sufficient annotations on the target application. Several approaches dedicated to vascular segmentation were proposed [23,28,33]. More recently, approaches were developed to improve the vascular connectivity of the segmentation results. Classic vesselness filters were used to help the network model tubular shapes [24,30]. Alternative approaches focused on adapting the segmentation architecture to facilitate the learning of a function that preserves connectivity. Attention modules [35] were incorporated into architectures such as U-Net [23], and a topology-aware feature synthesis network was proposed to correct the prediction topology based on the Euler characteristic [16]. Proxy tasks were also introduced to help the model focus on the structure topology such as the centerline extraction or distance-map computation [14]. Many works proposed dedicated loss functions to improve the result connectivity [6,9,18,25,26,31]. All these connectivity-preserving strategies assume that a large annotated dataset is available, which is rarely the case in vascular imaging applications.

Another research direction was explored consisting of the design of post-processing techniques dedicated to the reconnection of vascular segmentation results. Various algorithms have been suggested, relying on centerlines [7], graphs [11,22], and contour completion processes [36]. These approaches are complex to use due to their high dependence on parameter selection, and none of them provide the code necessary to reproduce or compare their results.

In [2], they proposed a strategy to train the reconnecting model and used it to develop an unsupervised plug-and-play segmentation approach. We recognized the potential for the reconnecting term to be applied more broadly as a post-processing step for any type of vascular segmentation result. In the present article, we thoroughly investigate this idea.

In this article, we conduct a comprehensive investigation of this concept. We specifically examine the properties of the reconnecting term, focusing on the impact of the disconnection size parameter and its convergence behavior. Furthermore, we demonstrate the practicality and versatility of this new

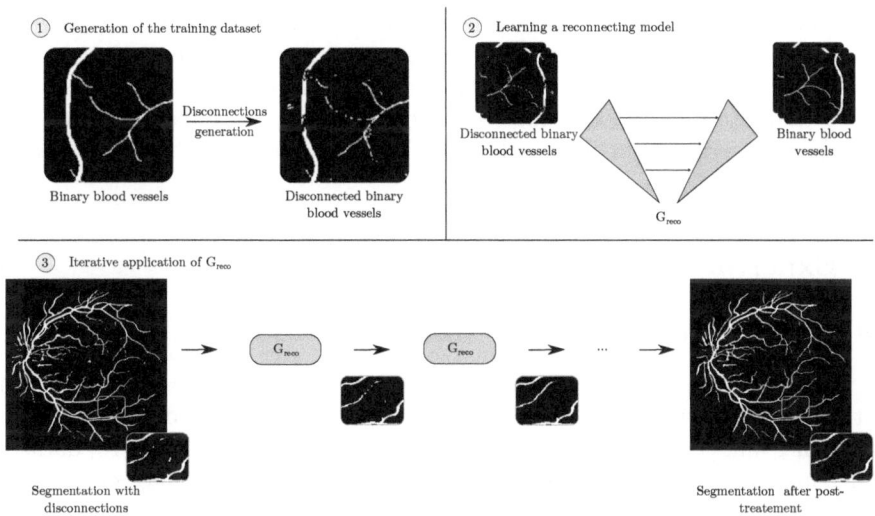

Fig. 1. Pipeline of our method. (1) a dataset is generated containing pairs of connected and disconnected vascular structures. (2) this dataset is used to train a model G_{reco} with a residual U-Net architecture. (3) finally, the trained model is iteratively applied on vascular segmentations with disconnections.

post-processing strategy by applying it to outputs from various segmentation methods and datasets, and comparing it with a recently developed reconnecting post-processing approach.

2   Proposed Method

In [2], a model, G_{reco}, based on a residual U-Net [13], is learned to reconnect disconnected vessel-like structures from a binary segmentation result. This model is trained on pairs of images containing connected and disconnected vessel-like structures (see top of Fig. 1).

The reconnecting term can be learned either based on manual annotations from the dataset of interest, or solely from synthetic images. This approach eliminates the need for an annotated dataset while still producing satisfactory results. When using synthetic images for training, an algorithm capable of generating random yet realistic disconnections from any binary vascular structure was proposed. To control the reconnection power of the model, disconnections are generated with various sizes drawn from a Gaussian distribution with mean s and standard deviation σ.

In this work, we propose to use this reconnecting model as a standalone post-processing step applicable to any binary segmentation result.

Intuitively, the size of the disconnections s of the training dataset should be adjusted to match the disconnection sizes in the segmentation. A small value may be insufficient to reconnect a vessel with a large gap, whereas a large value increases the risk of false reconnections. To address both small and large disconnections while minimizing false reconnections, we propose applying G_{reco} iteratively with a small value of s. This approach reconnects vessels gradually rather than attempting long-range reconnections in a single step (see bottom of Fig. 1). The code of our approach is available at https://github.com/creatis-myriad/plug-and-play-reco-regularization.

3   Experiments

3.1   Experimental Set up

We tested our framework in both 2D and 3D and used both synthetic and real datasets. In 2D, we used the DRIVE [32] dataset composed of 40 retinophotographies and their manual vascular annotations, and the STARE [10] dataset, which includes 20 annotated retinophotographies. The STARE dataset was used to train our reconnecting model, while the DRIVE dataset served as the test set for applying G_{reco}. Additionally, we generated a synthetic dataset comprising 20 synthetic vascular trees with OpenCCO [12], which was also employed as a training dataset for G_{reco}. In 3D, we used the Bullitt [1] and IXI[1] [34] datasets composed of 33 and 22 brain magnetic resonance angiographies (MRA) respectively and their manual vascular annotations. IXI served to train our reconnecting model while Bullitt was used as a test set to apply G_{reco}.

The disconnection algorithm was used on the DRIVE, STARE and IXI datasets to generate disconnected vascular trees with several mean disconnection sizes ($s \in 6, 8, 10, 12$). We experimentally set the standard deviation $\sigma = 4$ for DRIVE and STARE and to $\sigma = 2$ for IXI.

The backbone architecture for G_{reco} is a residual U-Net model [13] trained for 1000 epochs in 2D and 3000 epochs in 3D. We used an Adam optimizer with a learning rate of 10^{-3}. We employed a weighted Dice loss function as presented in [2] and a batch size of 32 in 2D and 4 in 3D. In 2D, the models were trained with an 80% split for training and 20% for validation, while in 3D, the split was 90% for training and 10% for validation. The final model was selected based on the best validation loss achieved during the training.

We selected three metrics to evaluate the segmentations. The classic Dice coefficient (DSC) assesses the overall quality of the segmentation. The Average Symmetric Surface Distance (ASSD) evaluates the segmentation geometry by measuring the distances between each element of the segmentation contour and the corresponding contour in the annotation. Finally, the error ratio of the number of connected components ϵ_{β_0} evaluates the segmentation connectivity. This error ratio is defined as $\epsilon_{\beta_0} = \left| \frac{\beta_0 - \beta_{0gt}}{\beta_{0gt}} \right|$, with β_0 the number of connected components of the segmentation and β_{0gt} the number of connected component of the annotation. The error ratio was preferred over the value of β_0 as β_0 is usually larger than 1 in the DRIVE and Bullitt annotations.

---

[1] https://brain-development.org/ixi-dataset/.

3.2   Ablation Study

Influence of the Size of the Disconnections.   The training dataset is a key element of our framework as it defines the concept of reconnection. The main parameter of this training dataset is the mean size of a disconnection, denoted as s. We trained four different models, denoted $G_{reco,s=X}$ (with $X \in 6, 8, 10, 12$), on the OpenCCO dataset which has been disconnected with a mean disconnection size X. We then tested these 4 models on the annotations of the DRIVE dataset that have also been disconnected with increasing values of s ($s = 6, s = 8, s = 10$ or $s = 12$). Hence each model will be tested on disconnection sizes it has not been trained for. In this experiment, G_{reco} has only been applied once to the segmentation results. The results are presented in Table 1.

Table 1. Results of applying G_{reco} trained on the OpenCCO dataset with several values of s, and applied on the Drive dataset with several values of s.

Training \ Test	$s = 6$			$s = 8$			$s = 10$			$s = 12$		
	DSC ↑	ASSD ↓	ϵ_{β_0} ↓	DSC ↑	ASSD ↓	ϵ_{β_0} ↓	DSC ↑	ASSD ↓	ϵ_{β_0} ↓	DSC ↑	ASSD ↓	ϵ_{β_0} ↓
Before G_{reco}	0.979	0.202	96.811	0.974	0.22	107.367	0.97	0.232	122.198	0.963	0.243	132.617
	± 0.004	± 0.06	± 67.065	± 0.004	± 0.071	± 71.883	± 0.005	± 0.057	± 83.665	± 0.007	± 0.054	± 86.553
$G_{reco,s=6}$	0.983	0.074	**11.429**	0.98	0.085	**14.095**	0.978	0.101	**15.485**	0.974	0.116	**17.619**
	± 0.003	± 0.012	± 8.783	± 0.003	± 0.015	± 10.92	± 0.004	± 0.019	± 10.758	± 0.004	± 0.022	± 13.8
$G_{reco,s=8}$	**0.985**	**0.067**	15.461	**0.983**	**0.077**	17.301	**0.981**	**0.09**	19.365	**0.977**	**0.103**	23.039
	± 0.002	± 0.013	± 11.256	± 0.003	± 0.014	± 12.69	± 0.004	± 0.019	± 14.269	± 0.004	± 0.02	± 18.927
$G_{reco,s=10}$	0.984	0.078	14.61	0.981	0.089	16.829	0.979	0.101	18.82	0.975	0.117	22.056
	± 0.003	± 0.015	± 11.223	± 0.003	± 0.016	± 13.051	± 0.004	± 0.023	± 14.516	± 0.004	± 0.017	± 17.864
$G_{reco,s=12}$	0.982	0.089	16.232	0.98	0.102	18.561	0.977	0.118	19.776	0.974	0.126	20.612
	± 0.003	± 0.015	± 12.213	± 0.003	± 0.017	± 13.75	± 0.004	± 0.024	± 14.034	± 0.004	± 0.019	± 16.189

As expected, the Dice coefficient of all models does not show a significant improvement after applying our post-processing, as the reconnection fragments represent only a small portion of the overall vessels. However both ϵ_{β_0} and $ASSD$ significantly decrease compared to the initial segmentation (Before G_{reco}), indicating that our post-processing successfully reconnected fragments of vessels.

We also observe a correlation between the size of the disconnections in the training dataset and the reconnections that occur. Models trained on larger disconnections tend to perform better on large disconnections (see Fig. 1 in Supplementary materials), while still facing challenges in reconnecting smaller disconnections. The model trained with disconnections with a size of $s = 8$, seems a good compromise, effectively reconnecting vessels while minimizing excessive false connections.

Convergence of the Proposed Approach.   In the previous experiment, we evaluated our models after a single application to better understand their behavior. However, our goal is to use G_{reco} iteratively to address disconnections that can not be fully reconnected in a single iteration. In this experiment, we analyze the interest of this iterative approach. Additionally, given the limited number of disconnections in an image, we anticipate that applying G_{reco} iteratively will

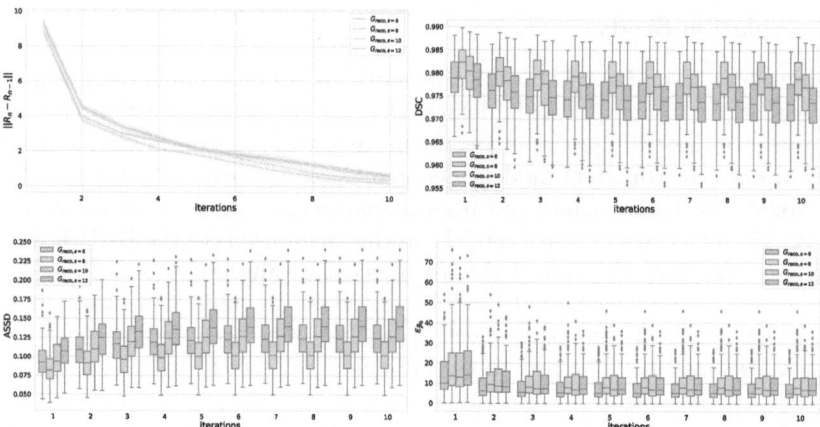

Fig. 2. Results of applying G_{reco}, trained on OpenCCO, on the DRIVE annotations artificially disconnected with sizes $s \in \{6, 8, 10, 12\}$. (a) convergence curves displaying the ℓ_2 norm of the difference of the last two consecutive results. (b–d) quantitative results of our models applied with an increasing number of iterations.

converge to a fixed-point image where all disconnections are filled. Thus, we also explore the experimental convergence of our framework in this experiment.

We used the same 4 reconnecting models $G_{reco,s=X}$ ($X \in 6, 8, 10, 12$) and applied each one on the Drive dataset which has been disconnected with several mean disconnection sizes ($s \in 6, 8, 10, 12$). The results are presented in Fig. 2.

We observe that the iterative application of G_{reco} converges as shown in Fig. 2(a). Interestingly this convergence occurs even though G_{reco} was applied to images with disconnection sizes different from those used in the training set. This highlights the robustness of our approach.

Figures 2 (b–d) show that the size of the disconnections in the training dataset has an impact on the quality of the reconnections made on the segmentation results. The model trained on the dataset created with the parameter $s = 8$ appeared to be a good compromise between significant reconnections (small ϵ_{β_0}) and limiting false reconnections (small $ASSD$ and high DSC values). Qualitative results are discussed in Supplementary materials.

3.3   Applications

In this experiment, we applied our post-processing to real 2D and 3D segmentation results obtained from both an unsupervised variational approach and a supervised deep learning approach. Specifically, we used the variational segmentation method proposed by Chan *et al.* [4] with a total variation [27] regularization. A gold standard U-Net architecture was used as the supervised approach as detailed in [2]. The variational segmentation yields disconnected and noisy results whereas the supervised one produces more complete and connected segmentations. We chose to evaluate our framework on these two different types of results to highlight the versatility of our post-processing method.

Table 2. Quantitative results obtained with our 2D and 3D reconnecting models on variational and deep learning segmentations. The p-values (from the t-test for normal distributions, or Wilcoxon test otherwise) are shown between the segmentation and $G_{reco, STARE}$ in 2D and the segmentation and $G_{reco_{IXI}}$ in 3D.

	training \ Test	Variational approach			Deep Learning		
		DSC	ASSD	ϵ_{β_0}	DSC	ASSD	ϵ_{β_0}
2D	Segmentation	0.758 ± 0.025	2.017 ± 0.452	98.543 ± 91.876	**0.811** ± 0.015	**1.155** ± 0.181	34.037 ± 26.858
	PRW [22]	0.758 ± 0.025	**2.012** ± 0.448	88.909 ± 85.873	0.808 ± 0.015	1.171 ± 0.177	22.017 ± 18.588
	$G_{reco,CCO}$	0.767 ± 0.023	2.423 ± 0.582	9.003 ± 11.388	0.809 ± 0.016	1.192 ± 0.19	**9.111** ± 7.606
	$G_{reco, STARE}$	**0.768** ± 0.023	2.332 ± 0.533	**6.609** ± 5.817	0.810 ± 0.015	1.198 ± 0.183	11.374 ± 9.095
	p-values	$\sim 10^{-6}$	0.057	$\sim 10^{-6}$	0.832	0.466	$\sim 10^{-6}$
3D	Segmentation	0.476 ± 0.02	**3.587** ± 0.42	26.562 ± 10.264	**0.756** ± 0.015	**1.488** ± 0.211	3.203 ± 1.689
	$G_{reco, IXI}$	**0.495** ± 0.019	4.154 ± 0.427	**5.618** ± 2.411	0.74 ± 0.014	1.552 ± 0.21	**1.697** ± 1.029
	p-values	$\sim 10^{-4}$	$\sim 10^{-6}$	$\sim 10^{-17}$	$\sim 10^{-5}$	0.226	$\sim 10^{-10}$

In 2D, we trained our reconnecting model G_{reco} either on the synthetic OpenCCO dataset, denoted as $G_{reco,CCO}$, or the STARE dataset, denoted as $G_{reco,STARE}$, that have both been disconnected with a mean disconnection size set to $s = 8$. We ran both segmentation strategies (variational and deep-learning) on the DRIVE dataset and applied our post-processing.

In 3D, we trained our reconnecting model on the IXI dataset, denoted as $G_{reco,IXI}$, that have been disconnected with a mean disconnection size set to $s = 8$. We ran both segmentation strategies on the Bullitt dataset and applied our post-processing.

To our knowledge, no other reconnecting post-processing approach offers open-access code. Consequently, we re-implemented the probability regularized walk (PRW) algorithm from [22] for comparison with our results. This method integrates the probability output of a neural network and the directions of broken vessel segments into a regularized walk algorithm to reconnect them to the main component. The hyperparameters of this method have been optimized for each image using a grid search. Due to incomplete details in the original article [22], we made certain choices that might differ from those of the authors. These decisions were made to ensure a fair comparison. Our implementation can be accessed at https://github.com/creatis-myriad/plug-and-play-reco-regularization. Results are summarized in Table 2, Fig. 3.

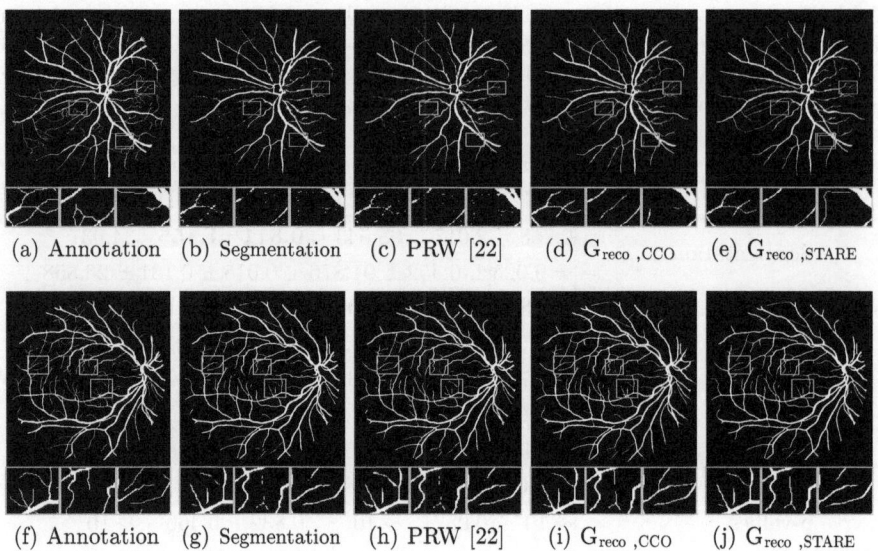

(a) Annotation   (b) Segmentation   (c) PRW [22]   (d) G_{reco} ,CCO   (e) G_{reco} ,STARE

(f) Annotation   (g) Segmentation   (h) PRW [22]   (i) G_{reco} ,CCO   (j) G_{reco} ,STARE

Fig. 3. Results of our post-processing applied to a DRIVE segmentation result from the variational (top row) and deep learning approach (bottom row). (b) and (g) depict segmentation results from variational and deep learning approaches respectively, before post-processing with G_{reco}. G_{reco} is trained on either the synthetic OpenCCO dataset (d) and (i) or the real STARE dataset (e) and (j). (c) and (h) represent the post-processing proposed in [22].

We observe that, in general, our post-processing either slightly increases the DSC and ASSD values of the segmentations or does not significantly change them (p-value < 0.05). The slight decrease in the ASSD of the 3D variational approach and the DSC of the 3D deep learning approach can be attributed to some false reconnection of aligned artifacts. The overall stability of the DSC and ASSD metrics is expected, because our post-processing primarily involves adding a few pixels to reconnect vessels as discussed before. By contrast, we observe a drastic decrease in the connectivity metric ϵ_{β_0}, since our post-processing affects the connectivity of the segmentations. Specifically, we note a decrease of $>90\%$ in 2D, and $>80\%$ in 3D for the variational approach, and $>67\%$ in 2D, and $>47\%$ in 3D for the deep learning approach.

It is interesting to note that the model trained on the synthetic OpenCCO dataset yields slightly less inferior results, often due to the creation of false connections. Our reconnecting terms learn to reconnect only based on geometric features. Therefore, the closer the geometry of the vessels in the training dataset to that of the test dataset, the better the performance tends to be. Nonetheless, the drop of performance is quite small which makes our term very useful in a purely unsupervised context in 2D when no vascular annotation is available. In 3D, to the best of our knowledge, there is no synthetic vascular network generation software that yields vascular trees geometrically close enough to a real brain vascular network. In particular, the tortuosity of brain vessels is much higher than what is possible to generate with softwares like VascuSynth[2] or OpenCCO.

The PRW algorithm reconnects some vessels, as evidenced by the small decrease in ϵ_{β_0}, but it is significantly less effective than our approach. PRW relies on probability maps to determine the direction of reconnection. However, we observed that the probability maps generated by both the variational and deep learning approaches exhibit a low dynamic range and are nearly binary. Consequently, the reconnection capability of PRW is limited on these types of segmentation results. In contrast, our approach directly relies on binary images to predict the reconnecting direction.

Additionally, qualitatively, PRW tends to produce more unrealistic reconnections, such as vessel enlargements, compared to those generated by our method.

4   Conclusion

In this article, we introduced a novel vascular segmentation post-processing to favor vascular network connectivity. This post-processing can be used in an unsupervised or supervised context depending on the availability of vascular annotations on the dataset of interest. We conducted an extensive validation of our approach both in 2D and 3D and showed that our post-processing is robust to the size of disconnections, converges to a reconnected result when used iteratively, and significantly improves the connectivity of segmentation results. We also compared our approach to a recent reconnecting post-processing algorithm, demonstrating that our method reconnects significantly more vessels and does so in a more realistic manner. However, our approach is purely based on the vessel's geometric properties in binary segmentations and thus some false reconnections may appear. Future works include taking into account the intensities of the underlying image to avoid these false reconnections.

Acknowledgments. This work was supported by Agence Nationale de la Recherche (ANR-22-CE45-0018, ANR-18-CE45-0018), LABEX PRIMES (ANR-11-LABX-0063). This work was granted access to the HPC resources of IDRIS under the allocations 2022-AD011013887 and 2023-AD011014452 made by GENCI.

Disclosure of Interests. The authors have no competing interests to declare that are relevant to the content of this article.

References

1. Bullitt, E., et al.: Vessel tortuosity and brain tumor malignancy: a blinded study. Acad. Radiol. **12**, 1232–1240 (2005)
2. Carneiro-Esteves, S., et al.: A plug-and-play framework for curvilinear structure segmentation based on a learned reconnecting regularization. Neurocomputing **599**, 128055 (2024)

---

[2] https://vascusynth.cs.sfu.ca/Welcome.html.

3. Carrillo, J.F., et al.: Recursive tracking of vascular tree axes in 3d medical images. Int. J. Comput. Assist. Radiol. Surg. **1**, 331–339 (2007)

4. Chan, T., et al.: Algorithms for finding global minimizers of image segmentation and denoising models. SIAM J. Appl. Math. **66**(5), 1632–1648 (2006)

5. Chung, M., et al.: Accurate liver vessel segmentation via active contour model with dense vessel candidates. Comput. Methods Programs Biomed. **166**, 61–75 (2018)

6. Clough, J.R., et al.: A topological loss function for deep-learning based image segmentation using persistent homology. IEEE Trans. Pattern Anal. Mach. Intell. **44**(12), 8766–8778 (2020)

7. Du, H., et al.: Retinal blood vessel segmentation by using the MS-LSDNet network and geometric skeleton reconnection method. Comput. Biol. Med. **153**, 106416 (2023)

8. Frangi, A.F., Niessen, W.J., Vincken, K.L., Viergever, M.A.: Multiscale vessel enhancement filtering. In: Wells, W.M., Colchester, A., Delp, S. (eds.) MICCAI 1998. LNCS, vol. 1496, pp. 130–137. Springer, Heidelberg (1998). https://doi.org/10.1007/BFb0056195

9. Hakim, L., et al.: Regularizer based on Euler characteristic for retinal blood vessel segmentation. Pattern Recogn. Lett. **149**, 83–90 (2021)

10. Hoover, A., et al.: Locating blood vessels in retinal images by piecewise threshold probing of a matched filter response. IEEE Trans. Med. Imaging **19**(3), 203–210 (2000)

11. Joshi, V.S., et al.: Identification and reconnection of interrupted vessels in retinal vessel segmentation. In: 2011 IEEE International Symposium on Biomedical Imaging: From Nano to Macro, pp. 1416–1420. IEEE (2011)

12. Kerautret, B., et al.: OpenCCO: an implementation of constrained constructive optimization for generating 2D and 3D vascular trees. Image Process. On Line **13**, 258–279 (2023)

13. Kerfoot, E., et al.: Left-ventricle quantification using residual U-Net. In: International Workshop on Statistical Atlases and Computational Models of the Heart, pp. 371–380 (2018)

14. Keshwani, D., Kitamura, Y., Ihara, S., Iizuka, S., Simo-Serra, E.: TopNet: topology preserving metric learning for vessel tree reconstruction and labelling. In: Martel, A.L. (ed.) MICCAI 2020, Part VI. LNCS, vol. 12266, pp. 14–23. Springer, Cham (2020). https://doi.org/10.1007/978-3-030-59725-2_2

15. Lamy, J., et al.: A benchmark framework for multi-region analysis of vesselness filters. IEEE Trans. Med. Imaging **41**, 3649–3662 (2022)

16. Li, L., et al.: Robust segmentation via topology violation detection and feature synthesis. In: Greenspan, H., et al. (eds.) MICCAI 2023. LNCS, vol. 14223, pp. 67–77. Springer, Cham (2023). https://doi.org/10.1007/978-3-031-43901-8_7

17. Liao, W., et al.: Progressive minimal path method for segmentation of 2D and 3D line structures. IEEE Trans. Pattern Anal. Mach. Intell. **40**(3), 696–709 (2017)

18. Lin, M., et al.: DTU-Net: learning topological similarity for curvilinear structure segmentation. In: Frangi, A., de Bruijne, M., Wassermann, D., Navab, N. (eds.) IPMI 2023. LNCS, vol. 13939, pp. 654–666. Springer, Cham (2023). https://doi.org/10.1007/978-3-031-34048-2_50

19. Merveille, O., et al.: Curvilinear structure analysis by ranking the orientation responses of path operators. IEEE Trans. Pattern Anal. Mach. Intell. (PAMI) **40**(2), 304–317 (2018)

20. Merveille, O., et al.: n d variational restoration of curvilinear structures with prior-based directional regularization. IEEE Trans. Image Process. **28**(8), 3848–3859 (2019)

21. Miraucourt, O., et al.: Variational method combined with Frangi vesselness for tubular object segmentation. In: Computational & Mathematical Biomedical Engineering (CMBE), pp. 485–488 (2015)

22. Mou, L., et al.: Dense dilated network with probability regularized walk for vessel detection. IEEE Trans. Med. Imaging 39(5), 1392–1403 (2020)

23. Mou, L., et al.: CS2-Net: deep learning segmentation of curvilinear structures in medical imaging. Med. Image Anal. 67, 101874 (2021)

24. Peng, Y., et al.: Curvilinear object segmentation in medical images based on ODoS filter and deep learning network. *arXiv preprint* arXiv:2301.07475 (2023)

25. Qiu, Y., et al.: CorSegRec: a topology-preserving scheme for extracting fully-connected coronary arteries from CT angiography. In: Greenspan, H., et al. (eds.) MICCAI 2023. LNCS, vol. 14222, pp. 670–680. Springer, Cham (2023). https://doi.org/10.1007/978-3-031-43898-1_64

26. Rougé, P., et al.: Cascaded multitask U-Net using topological loss for vessel segmentation and centerline extraction (2023)

27. Rudin, L.I., et al.: Nonlinear total variation based noise removal algorithms. Physica D 60(1–4), 259–268 (1992)

28. Sanchesa, P., et al.: Cerebrovascular network segmentation of MRA images with deep learning. In: 2019 IEEE 16th International Symposium on Biomedical Imaging (ISBI 2019), pp. 768–771. IEEE (2019)

29. Sato, Y., et al.: Three-dimensional multi-scale line filter for segmentation and visualization of curvilinear structures in medical images. Med. Image Anal. 2(2), 143–168 (1998)

30. Shi, T., et al.: Local intensity order transformation for robust curvilinear object segmentation. IEEE Trans. Image Process. 31, 2557–2569 (2022)

31. Shit, S., et al.: clDice-a novel topology-preserving loss function for tubular structure segmentation. In: Proceedings of the IEEE/CVF Conference on Computer Vision and Pattern Recognition, pp. 16560–16569 (2021)

32. Staal, J., et al.: Ridge-based vessel segmentation in color images of the retina. IEEE Trans. Med. Imaging 23(4), 501–509 (2004)

33. Tetteh, G., et al.: DeepVesselNet: vessel segmentation, centerline prediction, and bifurcation detection in 3-d angiographic volumes. Front. Neurosci. 14, 1285 (2020)

34. Valderrama, N.: JoB-VS: joint brain-vessel segmentation in TOF-MRA images. In: IEEE International Symposium on Biomedical Imaging, ISBI 2023, Cartagena de Indias, Colombia, Cartagena de Indias, 18–21 April 2023 (2023)

35. Vaswani, A., et al.: Attention is all you need. Adv. Neural Inf. Process. Syst. 30 (2017)

36. Zhang, J., et al.: Reconnection of interrupted curvilinear structures via cortically inspired completion for ophthalmologic images. IEEE Trans. Biomed. Eng. 65(5), 1151–1165 (2018)

Multi-factor Component Tree Loss Function: A Topology-Preserving Method for Skeleton Segmentation from Bone Scintigrams

Anh Q. Nguyen[1]([✉]), Jean Cousty[2], Yukiko Kenmochi[3],
Shigeaki Higashiyama[4], Joji Kawabe[4], and Akinobu Shimizu[1]

[1] Department of Electrical Engineering and Computer Science,
Graduate School of Engineering, Tokyo University of Agriculture and Technology,
Tokyo, Japan
nguyenquynhanh1804a@gmail.com

[2] LIGM, Université Gustave Eiffel, CNRS, Marne-la-Vallée, France

[3] Normandie Université, UNICAEN, ENSICAEN, CNRS, GREYC, Caen, France

[4] Graduate School of Medicine, Osaka Metropolitan University, Osaka, Japan

Abstract. Accurate skeleton segmentation of the entire anteroposterior bone scintigrams of the human body is essential for diagnosing bone metastases. However, conventional methods lack a loss design incorporating prior anatomical information, leading to segmentation failures, particularly when dealing with the irregular shapes of organs or high concentrations of positive accumulation. Cases where diagnostic support systems present anatomically abnormal findings may shatter the confidence of doctors and their reliability in these systems. In this paper, we propose a novel multi-factor component tree loss function to resolve the topological issues in segmentation failures. The proposed loss function, computed based on the component trees, comprises two factors: image maxima vanishment and reconnection. We aim to discard the false positive connected components (FPCCs) and reconnect the disconnected true positive connected components (TPCCs) for each bone. Experiments conducted on a private bone scintigrams dataset show that our proposed method outperforms state-of-the-art approaches in dice similarity coefficient (DSC) while efficiently addressing topological issues at a low computational cost. Code is available at https://github.com/MultiCTree/MultiCTree.

Keywords: Skeleton segmentation · Bone scintigrams · Component tree

1 Introduction

Prostate and breast cancer are the two most common cancers diagnosed in men and women, respectively, often metastasizing to the bones [1]. Bone scintigra-

Supplementary Information The online version contains supplementary material available at https://doi.org/10.1007/978-3-031-73967-5_7.

phy, a nuclear medicine procedure, is commonly used for diagnostic purposes [2]. Bone scintigraphy produces two-dimensional images called bone scintigrams, including each patient's anterior and posterior sides. In bone scintigraphy, the Bone Scan Index (BSI) is valuable for quantitatively evaluating bone metastases spread [3]. Accurate skeleton segmentation in the anteroposterior images of the whole body is required for the calculation. Shimizu et al. [4] proposed a system for skeleton segmentation from bone scintigrams. However, many recognition failures occurred due to high concentrations of positive accumulation and irregular organ shapes. Yu et al. [5] compared three advanced CNN-based models for skeleton segmentation without prioritizing anatomical improvement or topological evaluation metrics. To integrate topological prior, Clough et al. [6] introduced a persistent homology loss function, which can be used for n-dimensional images, although generating persistent diagrams is time-consuming. Shit et al. [7] proposed a soft-clDice loss function based on a soft skeleton algorithm. It suits tubular structures like blood vessels rather than thick structures like bones. Perret et al. [8] designed a component tree loss function to reinforce or discard image maxima based on attributes. Still, it neither reconnects components nor accurately determines components to retain or discard based on the ground truths.

In this paper, we aim to improve the anatomical correctness of segmentation failures from bone scintigrams that reduce the confidence of medical professionals. Our major contribution is proposing a novel multi-factor component tree loss function to eliminate false positive connected components (FPCCs) and establish proper reconnections among disconnected true positive connected components (TPCCs). This loss function improves the reliability of skeleton segmentation methodologies, specifically addressing challenges related to anatomical aspects.

2   Method

2.1   Max-Tree

A grayscale image can be represented by a component tree, which contains all necessary information about the image components and the relationship between each component at each level of the level sets [9]. In a component tree, each connected component serves as a node, and the inclusion relationships between the connected components form the edges [8].

Let $V = \{v_i\}_{i \in [1,n]}$ denote the finite nonempty set made with n image pixels. An image is represented as a vector $\mathbf{f} \in \mathbb{R}^n$, and for any $i \in [1,n]$, $\mathbf{f}_i$ is the value of pixel v_i. For any $\lambda \in \mathbb{R}$, $[\mathbf{f}]_\lambda$ is the level set of $\mathbf{f}$ of level λ: $[\mathbf{f}]_\lambda = \{v_i \in V \mid \mathbf{f}_i \geq \lambda\}$. The *set of all connected components of* $[\mathbf{f}]_\lambda$ is denoted by $\mathcal{CC}([\mathbf{f}]_\lambda)$ and, for any λ in $\mathbb{R}$, any element in $\mathcal{CC}([\mathbf{f}]_\lambda)$ is also called a *connected component of* $\mathbf{f}$. The set of all connected components of $\mathbf{f}$ is written as $\mathcal{CC}(\mathbf{f})$, so that $\mathcal{CC}(\mathbf{f}) = \bigcup_{\lambda \in \mathbb{R}}\{\mathcal{CC}([\mathbf{f}]_\lambda)\}$. The set $\mathcal{CC}(\mathbf{f})$ is finite, and if the number of connected components of $\mathbf{f}$ is m, we can write $\mathcal{CC}(\mathbf{f}) = \{C_i\}_{i \in [1,m]}$ where C_i is the i-th connected component. The *altitude* of C_i is defined by $alt(C_i) = \max\{\lambda \in \mathbb{R} \mid C_i \in \mathcal{CC}([\mathbf{f}]_\lambda)\}$. The *max-tree* $\mathrm{MT}(\mathbf{f})$ of $\mathbf{f}$ is a pair $(\mathrm{MT}_1, \mathrm{MT}_2)$ made of two parts: $\mathrm{MT}_1(\mathbf{f})$ represents the connected components, and $\mathrm{MT}_2(\mathbf{f})$ is the altitude vector of the components. Thus, we have $(\mathrm{MT}_1(\mathbf{f}), \mathrm{MT}_2(\mathbf{f})) = (\{C_i\}_{i \in [1,m]}, \mathbf{a})$, where the i-th element of vector $\mathbf{a}$ in $\mathbb{R}^m$ is $alt(C_i)$ [8].

Figure 1 shows an example of a one-dimensional image and Fig. 2 shows its max-tree. In Fig. 2, orange and green-filled rectangles represent max-tree nodes including the maxima with their altitudes. The nodes are linked in a parent-child relation (highlighted with solid lines) derived from the inclusion between the connected components. A node without a parent is called a *root*, and one without a child is a *leaf*, also called a *(regional) maximum*.

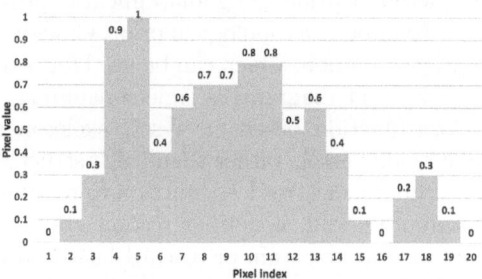

Fig. 1. One-dimensional image with pixel index and pixel value.

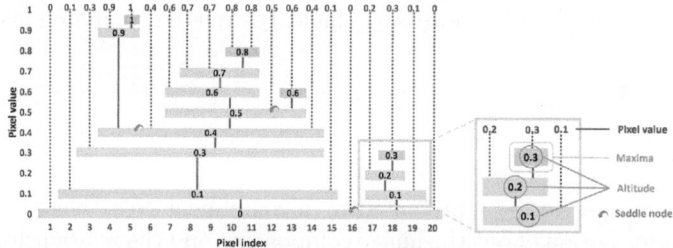

Fig. 2. Max-tree of the one-dimensional image shown in Fig. 1.

2.2 Original Component Tree Loss Function

Assuming k is the number of maxima in the hierarchy of $\mathrm{MT}(\mathbf{f})$, ℓ is a target maxima number, and μ in $\mathbb{R}$ is a constant margin to prevent selected maxima growing without limit. The vectors $\mathbf{sm} \in \mathbb{R}^k$ and $\mathbf{im} \in \mathbb{R}^k$ represent saliency and importance measures on the maxima, respectively. Importance measures rank the maxima for reinforcement or discarding while adjusting saliency measures

help reinforce or discard maxima [8]. The original component tree loss function
is as follows:

$$\mathcal{L}_{\text{CTree}} = \sum_{i=1}^{i\leq\ell} \max(\mu - \mathbf{sm}_{\mathbf{r}_i}, 0) + \sum_{i=\ell+1}^{i\leq k} \mathbf{sm}_{\mathbf{r}_i} \qquad (1)$$

where $\mathbf{r} = \text{argsort}(\mathbf{im})$ is a permutation vector sorting the maxima indices of
$\mathbf{f}$ in descending order of importance measure. The first term of Eq. (1) rein-
forces ℓ target maxima while the second term discards the others based on
saliency measures. Importance measures such as *dynamics* [8] fail to distinguish
TPCCs and FPCCs while existing saliency measures cannot reconnect components. Hence, we improve the original component tree loss function by introducing and combining new saliency and importance measures into a novel loss
function.

2.3   Multi-factor Component Tree Loss Function

New Saliency Connect. In skeleton segmentation, disconnected components
within the same bone may occur. Strengthening the altitude of saddle nodes
between these components can address this issue. In the max-tree (Fig. 2), a
saddle node paired to a maximum M, denoted by $S(M)$, is the nearest ancestor
of M also containing a maximum with a higher altitude than the one of M. To
minimize the loss function using the saliency *dynamics* [8] (*i.e.*, *dynamics*$(M) =$
$alt(M) - alt(S(M))$, one has to simultaneously decreases $alt(M)$ and increases
$alt(S(M))$. However, to avoid reducing $alt(M)$ and purely establish reconnection,
we propose a novel saliency measure *connect*, to reconnect these disconnected
components. The saliency *connect* of M is obtained as the difference between a
constant value and the altitude of the saddle node $S(M)$ paired to M. Here, let
us set the constant value v to the maximum pixel value of the image. Then, we
have:

$$connect(M) = v - alt(S(M)) \qquad (2)$$

In Fig. 3, the saliency *dynamics* and *connect* of each maximum are represented by blue and green double arrows, along with their corresponding values.
In this case, because the maximum pixel value of the image is 1, v is set to 1.

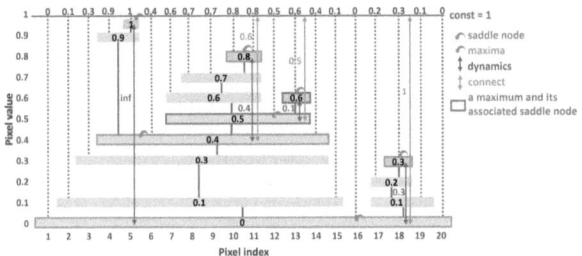

Fig. 3. Saliency measures *dynamics* and *connect*.

New Importance Precision. The attributes of FPCCs such as *dynamics* may be higher than those of TPCCs, resulting in erroneous reconnection paths with FPCCs [8]. Therefore, we propose a novel importance measure to categorize connected components as desirable TPCCs or FPCCs. Precision measures the ratio of correctly segmented pixels to all pixels identified as ground truth by the model. A TPCC is considered desirable if it has high precision. Let G be the set of ground truth pixels. The precision of a component C_i is defined as follows:

$$precision(C_i) = \frac{|C_i \cap G|}{|C_i|} \tag{3}$$

The importance of maxima is measured by an increasing attribute. The *extinction value* for an importance measure of a maximum M is the smallest threshold value θ at which M remains a maximum after removing all nodes with attribute values smaller than θ [8,10]. It can also be obtained as the importance measure of the saddle node $S(M)$ paired to M. Since *precision* is not increasing in the max-tree, we employed the max-rule regularization [11] on *precision* before computing the extinction value, denoted as Ext(*precision*(M)). See supplementary material for more details on the calculation of the extinction value.

Combination of New Saliency and Importance Measures. The new loss function integrates the proposed saliency *connect* and importance *precision*. All maxima are classified into two distinct groups: G_1 (for vanishment) and G_2 (for reconnection), as precisely defined in Table 1. Determining maxima in each group is based on the extinction value for the importance *precision* (the third column in Table 1). The *multi-factor component tree loss function* is defined as follows:

$$\mathcal{L}_{\text{multiCTree}} = \sum_{i \in G_1} \text{sm}_{1i} + \sum_{i \in G_2} \text{sm}_{2i} \tag{4}$$

where $\text{sm}_1 = alt$, and $\text{sm}_2 = connect$. It can be observed that to minimize $\mathcal{L}_{\text{multiCTree}}$, the maxima in G_1 should be eliminated by reducing their altitude, which removes FPCCs and undesirable TPCCs, while the maxima in G_2 (desirable TPCCs) should be reconnected using the saliency *connect* by increasing the altitude of their paired saddle nodes. We experimented with threshold θ values of 0.5 and 0.8 and determined that setting θ to 0.5 yields the best outcomes.

Table 1. Two groups of maxima: vanishment and reconnection.

Group	Saliency	Condition
G_1: Vanishment	*altitude*	Ext(*precision*(M_i)) $< \theta$
G_2: Reconnection	*connect*	Ext(*precision*(M_i)) $\geq \theta$

3   Experiments

3.1   Setup

Dataset. Experimental bone scintigrams dataset is a DICOM-format private dataset including 1,235 cases, totaling 2,470 images from the anterior and posterior sides of female patients. Images were taken with various devices and standardized to 576×256 pixels with a 2.8×2.8 mm/pixel resolution. Segmentation targets include 13 layers for 12 anterior bones (skull, cervical vertebrae, thoracic vertebrae, lumbar vertebrae, sacrum, pelvis, ribs, scapula, humerus, femur, sternum, clavicle)+background, and 14 layers for 10 posterior bones (skull, cervical vertebrae, thoracic vertebrae, lumbar vertebrae, sacrum, pelvis, ribs, scapula, humerus, femur, scapula∩ribs (scapula2), scapula∪scapula2 (scapula3), ribs∪scapula2 (ribs2))+background. Labels were annotated by the authors in [4].

Evaluation Metrics. Evaluation metrics include Dice similarity coefficient (DSC), the differences between predictions and ground truths for both (i) the number of connected components ($\Delta\#$(CCs)), and (ii) the number of TPCCs ($\Delta\#$(TPCCs)), the number of FPCCs ($\#$(FPCCs)), and the FPCC size of each bone. Connected components are defined by 8-connectivity. These topological metrics approaching zero show better topological segmentation results. Wilcoxon signed-rank test was used for the statistical comparison, with the null hypothesis: "There is no significant difference in the performance distributions between the two methods".

Loss Function. For consistent comparisons, the loss function of the model incorporating a topological loss function comprises three losses: cross-entropy loss $\mathcal{L}_{\mathrm{CE}}$, deep supervision loss $\mathcal{L}_{\mathrm{DSV}}$ [12] calculated based on dice loss, and a topological loss $\mathcal{L}_{\mathrm{topology}}$. Each loss is computed independently for anterior and posterior bones and then averaged. Equation (5) represents the total loss function. Note that the loss function of the base model is identical to Eq. (5) but without $\mathcal{L}_{\mathrm{topology}}$.

$$\mathcal{L}_{\mathrm{total}} = 0.5\mathcal{L}_{\mathrm{CE}} + 0.5\mathcal{L}_{\mathrm{DSV}} + \lambda\mathcal{L}_{\mathrm{topology}} \tag{5}$$

3.2   Implementation Details

Three-fold cross-validation was used, with data split 4:1:1 for training, validation, and testing. Models were trained for 50 epochs on NVIDIA A100 SXM4 80 GB, Python 3.8.10, and PyTorch 1.11.10+cu115. SGD optimizer [13] had a learning rate of 0.01 and a batch size of 2. Augmentation included a horizontal flip. Loss weights were 0.01 for persistent homology loss $\mathcal{L}_{\mathrm{PH}}$ and multi-factor component tree loss $\mathcal{L}_{\mathrm{multiCTree}}$, and 0.1 for soft-clDice loss $\mathcal{L}_{\mathrm{clDice}}$ and component tree loss $\mathcal{L}_{\mathrm{CTree}}$. These weight settings optimized the performance of each model.

3.3   Experiment 1

We evaluated the computational efficiency of our method against state-of-the-art (SOTA) methods: $\mathcal{L}_{PH}$ [6] (uses a topology layer [14]), $\mathcal{L}_{clDice}$ [7], and $\mathcal{L}_{CTree}$ [8]. For computational feasibility, topological loss functions were applied solely to anterior and posterior femurs, as the persistent diagram generation process for all bones is computationally expensive. Statistical tests compared the base model TransBtrflyNet [15] with and without topological loss functions. Null hypothesis is rejected at a significance level of 0.05(*) or 0.01(*). From Table 2, all the methods improved anterior DSC but reduced posterior DSC, potentially due to conflicts between loss functions within $\mathcal{L}_{total}$. In terms of topological metrics, except for $\mathcal{L}_{PH}$, the use of $\mathcal{L}_{clDice}$, $\mathcal{L}_{CTree}$, and $\mathcal{L}_{multiCTree}$ resulted in reductions. $\mathcal{L}_{PH}$ generated numerous #(FPCCs), leading to an increase in FPCC size, while $\mathcal{L}_{clDice}$ exhibited outstanding topological outcomes. $\mathcal{L}_{multiCTree}$ showed a superior anterior DSC and the best posterior Δ#(TPCCs) for femurs. For cases where the base model had topological issues, all topological loss functions improved results across all metrics. $\mathcal{L}_{PH}$ addressed all cases, while the others focused on cases having topological issues. See supplementary material for examples of difficult cases where images contain significant noise and numerous metastases appear on the bones. As shown in Table 3, $\mathcal{L}_{PH}$ shows the longest training time, while other methods are similar to the base model. Integrating topological loss functions does not significantly impact memory consumption. In conclusion, our methodology achieves near-optimal computational efficiency with comparable outcomes.

Table 2. Average results of femurs across 1235 cases (A: Anterior, P: Posterior; *:p < 0.05, **:p < 0.01; down arrow$^\downarrow$: inferiority, up arrow$^\uparrow$: superiority).

	Method	DSC$\uparrow$	Δ#(CCs)$\downarrow$	Δ#(TPCCs)$\downarrow$	#(FPCCs)$\downarrow$	FPCC size$\downarrow$
A	Base	0.9442	0.0348	0.0178	0.0170	0.9895
	$\mathcal{L}_{PH}$	0.9449**$^\uparrow$	0.0672**$^\downarrow$	0.0211	0.0462**$^\downarrow$	3.0057**$^\downarrow$
	$\mathcal{L}_{clDice}$	0.9446	**0.0154**$^{**\uparrow}$	**0.0097**$^{*\uparrow}$	**0.0057**$^{**\uparrow}$	**0.1474**$^{**\uparrow}$
	$\mathcal{L}_{CTree}$	**0.9454**$^{**\uparrow}$	0.0251	0.0138	0.0113	0.4356
	$\mathcal{L}_{multiCTree}(\theta = 0.5)$	0.9434**$^\uparrow$	0.0291	0.0162	0.0130	0.5134
P	Base	**0.9422**	0.0405	0.0227	0.0178	0.6283
	$\mathcal{L}_{PH}$	0.9404**$^\downarrow$	0.0575*$^\downarrow$	0.0308	0.0267	1.4219
	$\mathcal{L}_{clDice}$	0.9414**$^\downarrow$	**0.0235**$^{**\uparrow}$	0.0194	**0.0040**$^{**\uparrow}$	**0.1611**$^{*\uparrow}$
	$\mathcal{L}_{CTree}$	0.9414**$^\downarrow$	0.0364	0.0219	0.0146	0.5085
	$\mathcal{L}_{multiCTree}(\theta = 0.5)$	0.9415**$^\downarrow$	0.0324	**0.0178**$^{*\uparrow}$	0.0146	0.5668

Table 3. Comparison of computational efficiency for one fold over a single epoch.

Method	Base	$\mathcal{L}_{\text{PH}}$	$\mathcal{L}_{\text{clDice}}$	$\mathcal{L}_{\text{CTree}}$	$\mathcal{L}_{\text{multiCTree}}(\theta = 0.5)$
Training time (MM:SS)	08:01	26:47	08:58	08:11	08:12
Memory (/80 GB)	37.2	37.5	37.2	37.2	37.2

3.4 Experiment 2

We applied topological loss functions to all anterior and posterior bones, with the same statistical procedure as Experiment 1. Table 4 shows the ratio of superior/inferior results to total segmentation targets (25 layers, excluding background, as defined in Sect. 3.1) for both anterior and posterior sides. Higher superiority ratios reflect better performance, while higher inferiority ratios show decreased performance. $\mathcal{L}_{\text{clDice}}$ achieves the most superior results, while $\mathcal{L}_{\text{CTree}}$ has the fewest inferior results. Experiments with $\theta = 0.5$ and $\theta = 0.8$ either competed with or outperformed SOTA methods for DSC, $\Delta\#(\text{CCs})$, and $\Delta\#(\text{TPCCs})$ for both superiority and inferiority. Table 5 shows average results of all bones of all cases, where $\mathcal{L}_{\text{multiCTree}}(\theta = 0.5)$ exhibits the best results in terms of DSC, $\Delta\#(\text{CCs})$, and $\Delta\#(\text{TPCCs})$. Increasing θ to 0.8 decreased $\#(\text{FPCCs})$ and FPCC size. Our method effectively reconnected TPCCs but concurrently generated undesirable FPCCs. $\mathcal{L}_{\text{multiCTree}}$ loss relies on the ground truths to define TPCCs and FPCCs. However, we investigated that bones such as ribs or the scapula were incorrectly labeled based on prior anatomical knowledge, resulting in an incorrect number of connected components in the ground truths. In addition, the current threshold θ is not optimal enough to eliminate FPCCs.

Table 4. The ratio of superior/inferior results to the total number of segmentation targets excluding background (S($\uparrow$): Superiority, I($\downarrow$): Inferiority).

Method	DSC		$\Delta\#(\text{CCs})$		$\Delta\#(\text{TPCCs})$		$\#(\text{FPCCs})$		FPCC size	
	S($\uparrow$)	I($\downarrow$)	S($\uparrow$)	I($\downarrow$)	S($\uparrow$)	I($\downarrow$)	S($\uparrow$)	I($\downarrow$)	S($\uparrow$)	I($\downarrow$)
$\mathcal{L}_{\text{clDice}}$	0.28	0.36	**0.24**	**0.00**	0.16	0.04	**0.32**	**0.00**	**0.12**	0.08
$\mathcal{L}_{\text{CTree}}$	0.40	0.32	0.08	**0.00**	0.04	**0.00**	0.04	**0.00**	0.08	**0.04**
$\mathcal{L}_{\text{multiCTree}}(\theta = 0.5)$	**0.72**	**0.20**	0.20	**0.00**	0.16	**0.00**	0.16	0.08	0.08	0.12
$\mathcal{L}_{\text{multiCTree}}(\theta = 0.8)$	0.48	**0.20**	0.20	**0.00**	**0.20**	**0.00**	0.12	0.08	0.04	0.08

Figure 4 illustrates examples where $\mathcal{L}_{\text{multiCTree}}$ (fourth and fifth columns) outperformed SOTA approaches. The first row shows the capability of our method to reconnect TPCCs, while the second row highlights its ability to discard FPCCs.

Table 5. Average results of anterior and posterior bones across all cases.

Method	DSC↑	$\Delta\#$(CCs)↓	$\Delta\#$(TPCCs)↓	$\#$(FPCCs)↓	FPCC size↓
Base	0.9252	0.0690	0.0646	0.0207	0.2525
$\mathcal{L}_{\text{clDice}}$	0.9262	0.0645	0.0627	**0.0126**	**0.2436**
$\mathcal{L}_{\text{CTree}}$	0.9261	0.0669	0.0626	0.0217	0.3115
$\mathcal{L}_{\text{multiCTree}}(\theta = 0.5)$	**0.9266**	**0.0578**	**0.0528**	0.0245	0.3174
$\mathcal{L}_{\text{multiCTree}}(\theta = 0.8)$	0.9262	0.0620	0.0575	0.0214	0.2880

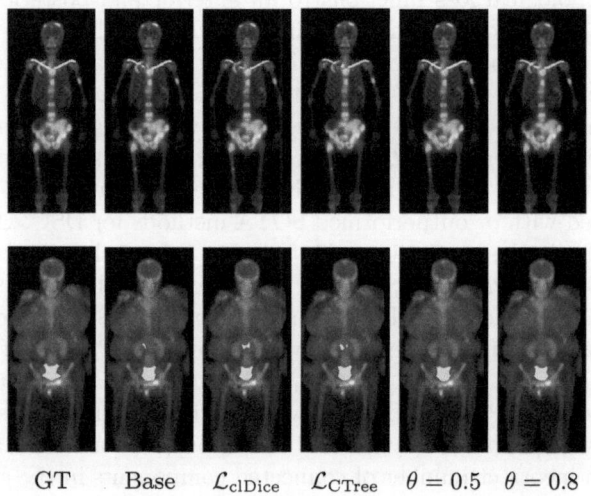

GT   Base $\mathcal{L}_{\text{clDice}}$ $\mathcal{L}_{\text{CTree}}$ $\theta = 0.5$ $\theta = 0.8$

Fig. 4. Comparison of different methods applied to all bones (GT: ground truth).

4 Conclusion

This paper introduces a novel multi-factor component tree loss function, integrating saliency *connect* and importance *precision*, to improve the topological accuracy of skeleton segmentation from bone scintigrams. For future work, we will correct ground truths and optimize the threshold θ for classifying maxima.

Acknowledgments. This study was supported by the JSPS Sakura program (JPJSBP 120233206) and PHC Sakura program (No. 49674K).

Disclosure of Interests. Akinobu Shimizu has received research grants from Nihon Medi-Physics Co., Ltd.

References

1. Siegel, R.L., Giaquinto, A.N., Jemal, A.: Cancer statistics, 2024. CA Cancer J. Clin. **74**(1), 8–9 (2024)
2. Donohoe, K.J., et al.: Appropriate use criteria for bone scintigraphy in prostate and breast cancer: summary and excerpts. J. Nucl. Med. **58**(4), 14N-17N (2017)
3. Imbriaco, M., et al.: A new parameter for measuring metastatic bone involvement by prostate cancer: the Bone Scan Index. Clin. Cancer Res. **4**(7), 1765–1772 (1998)
4. Shimizu, A., et al.: Automated measurement of bone scan index from a whole-body bone scintigram. Int. J. Comput. Assist. Radiol. Surg. **15**(3), 389–400 (2020)
5. Yu, P.N., Lai, Y.C., Chen, Y.Y., Cheng, D.C.: Skeleton Segmentation on Bone Scintigraphy for BSI Computation. Diagnostics (Basel, Switzerland) **13**(13), 2302 (2023)
6. Clough, J.R., Byrne, N., Oksuz, I., Zimmer, V.A., Schnabel, J.A., King, A.P.: A topological loss function for deep-learning based image segmentation using persistent homology. IEEE Trans. Pattern Anal. Mach. Intell. **44**(12), 8766–8778 (2022)
7. Shit, S., et al.: clDice - a novel topology-preserving loss function for tubular structure segmentation. In: Proceedings of the IEEE/CVF Conference on Computer Vision and Pattern Recognition (CVPR), pp. 16555–16564. IEEE Computer Society, United States (2021)
8. Perret, B., Cousty, J.: Component tree loss function: definition and optimization. In: Baudrier, É., Naegel, B., Krähenbühl, A., Tajine, M. (eds.) DGMM 2022. LNCS, vol. 13493, pp. 248–260. Springer, Cham (2022). https://doi.org/10.1007/978-3-031-19897-7_20
9. Souza, R., Tavares, L., Rittner, L. Lotufo, R.: An overview of max-tree principles, algorithms and applications. In: 2016 29th SIBGRAPI Conference on Graphics, Patterns and Images Tutorials (SIBGRAPI-T), Sao Paulo, Brazil, pp. 15–23 (2016)
10. Vachier, C., Meyer, F.: Extinction value: a new measurement of persistence. In: IEEE Workshop on Nonlinear Signal and Image Processing, pp. 254–257 (1995)
11. Salembier, P., Oliveras, A., Garrido, L.: Antiextensive connected operators for image and sequence processing. IEEE Trans. Image Process. Publ. IEEE Sig. Process. Soc. **7**(4), 555–570 (1998)
12. Dou, Q., et al.: 3D deeply supervised network for automated segmentation of volumetric medical images. Med. Image Anal. **41**, 40–54 (2021)
13. Robbins, H., Monro, S.: A stochastic approximation method. Herbert Robbins Selected Papers, pp. 102–109 (1985)
14. Gabrielsson, R.B., Nelson, B.J., Dwaraknath, A., Skraba, P.: A topology layer for machine learning. In: Proceedings of the Twenty Third International Conference on Artificial Intelligence and Statistics, pp. 1553–1563 (2020)
15. Nguyen, Q.A., Cousty, J., Kenmochi, Y., Higashiyama, S., Kawabe, J., Shimizu, A.: Improvement of a skeleton segmentation model of bone scintigrams with a transformer and component tree loss function. In: CARS 2023-Computer Assisted Radiology and Surgery Proceedings of the 37th International Congress and Exhibition Munich, Germany, 20–23 June 2023, Int J CARS, pp. S17–S18 (2023)

Exploitation of Mapper Algorithm in Neuroimaging Applications: A Novel Framework for Outcomes Prediction

Stefano Vannoni[1], Emma Tassi[1,2], Inês Won Sampaio[1],
and Eleonora Maggioni[1(✉)]

[1] Department of Electronics, Information and Bioengineering, Politecnico di Milano,
Milan, Italy
eleonora.maggioni@polimi.it
[2] Department of Neuroscience and Mental Health, Fondazione IRCCS Ca' Granda,
Osp. Maggiore Policlinico, Milan, Italy

Abstract. Topological Data Analysis (TDA) represents a pioneering methodology for revealing intricate structures within complex datasets. This study introduces a novel framework for leveraging the Mapper algorithm in neuroimaging studies. The proposed framework involves mapping new independent test samples onto a pre-constructed train graph, thereby harnessing embedded topological features to derive novel insights about test data. Validation of the framework employs a neuroimaging dataset sourced from the Human Connectome Project (HCP), encompassing white matter brain features, and includes practical applications for predicting categorical and continuous outcomes. The results validate the framework efficacy in transferring knowledge from train data to predict unseen samples, underscoring its potential across diverse neuroimaging applications.

This research highlights the potential of the Mapper-based TDA framework in neuroimaging, paving the way for its application across diverse neuroscience domains to extract clinically relevant features, improve predictive accuracy, and enhance patient treatment strategies. By discerning intricate patterns within high-dimensional patient data, this approach enables precise diagnostics and personalized treatment strategies, contributing to more accurate disease profiling and optimizing therapeutic interventions in personalized medicine.

Keywords: Topological Data Analysis · Mapper · Neuroimaging

1 Introduction

In recent years, Topological Data Analysis, colloquially referred as TDA, has emerged as a distinctive and novel approach offering a unique advantage point for comprehending the underlying structures embedded within complex datasets [14]. TDA facilitates the extraction of topological features from data,

C. Chen et al. (Eds.): TGI3 2024, LNCS 15239, pp. 76–87, 2025.
https://doi.org/10.1007/978-3-031-73967-5_8

thereby enabling the identification of significant patterns and relationships that might otherwise, remain concealed when employing traditional statistical techniques.

The two principal tools in TDA are persistent homology and Mapper algorithm. Persistent homology computes topological features of a space at varying spatial resolutions by representing it as a simplicial complex, thus identifying significant topological patterns across multiple scales. Mapper algorithm, developed by Singh et al. [8], effectively combines dimensionality reduction techniques, clustering algorithms, and graph networks to reveal data structures. It starts by applying a filter function $f : X \rightarrow R$ to transform high-dimensional data, influencing visualization and grouping. The algorithm then divides the filtered data into overlapping intervals, identifying points within each interval I_i, thus defining a cover of X. Clustering within each interval $(X_i \in I_i)$ produces X_{ik}, which form nodes in the graph. Nodes are connected by edges when $X_{ik} \cap X_{lm} \neq \emptyset$, indicating intersecting clusters. Figure 1 illustrates the schematic representation of the Mapper algorithm that has just been introduced.

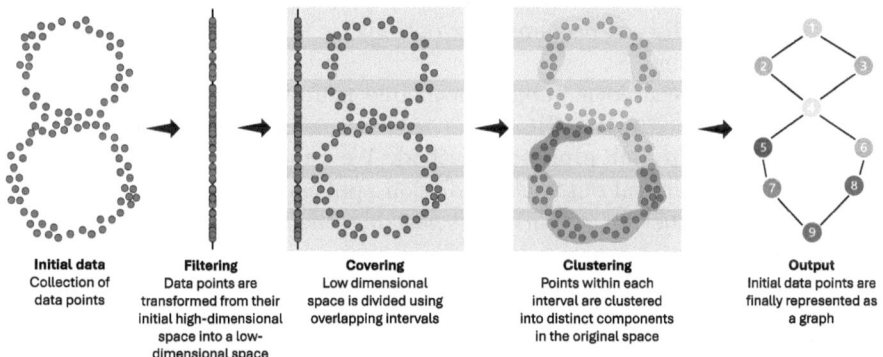

Initial data	Filtering	Covering	Clustering	Output
Collection of data points	Data points are transformed from their initial high-dimensional space into a low-dimensional space	Low dimensional space is divided using overlapping intervals	Points within each interval are clustered into distinct components in the original space	Initial data points are finally represented as a graph

Fig. 1. Schematic representation of the Mapper algorithm – The diagram illustrates the sequential steps of the Mapper algorithm for point cloud data analysis. The process begins with an initial collection of data points in a high-dimensional space. Subsequently, a filtering step is applied, transforming the data into a lower-dimensional space through filter function. The filtered space is then partitioned into overlapping intervals, creating a cover. Within each interval, clustering is performed on the corresponding points in the original space, identifying distinct components. The final output is a graph representation of the data, where nodes correspond to clusters and edges connect clusters with shared points across.

TDA has been successfully used in various fields such as time series analysis [4], medicine [9], biology [15], and chemistry [10], to mention a few. Our main focus is on the application of TDA, specifically leveraging the Mapper algorithm, in neuroimaging studies through prediction analyses. In typical neuroimaging applications [5,7], data are generally studied and analyzed individually, resulting in the construction of a separate graph for each input dataset. This approach frequently ignores the potential to infer new insights about independent data samples utilizing the information embedded within the constructed graphs.

In our work, we aim to introduce a novel framework that exploits the Mapper algorithm for mapping new independent samples onto a pre-constructed graph. The rationale behind this new framework is that a graph constructed using the Mapper algorithm encapsulates significant information about the input train data, making TDA particularly well-suited for applications in personalized medicine. Our methodology enables the analysis of new independent samples without necessitating the reconstruction of a new graph. Instead, it leverages the information embedded within a train graph to derive insights pertaining to these new samples. By projecting independent test samples onto this graph—specifically, mapping the test samples to the nodes of the graph—we aim to utilize the embedded information within these nodes to infer new insights about test data. In particular, this approach can yield significant insights into patient-specific anomalies and treatment responses, facilitating the integration of patient-specific data into graphs for tailored health profiles. By discerning intricate patterns within high-dimensional patient data, it enables precise diagnostics and the formulation of personalized treatment strategies, contributing to more accurate disease profiling and optimizing therapeutic interventions in personalized medicine.

To this end, we used a publicly available neuroimaging dataset from the Human Connectome Project (HCP) [12], which encompasses white matter brain features, to evaluate our proposed method. We validated the framework and established two practical outcome prediction applications: one for a categorical variable and the other for a continuous variable, thereby supporting the framework rationale.

2   Methods

For this study, the Mapper algorithm was implemented using KeplerMapper [13], a Python-based software library, while DyNeuSR [3], an open-source neuroinformatics platform, was utilized for visualization purposes. Statistical test were performed using SciPy, an open-source Python library.

2.1   Mapper Algorithm Parameters Setting

Regarding the three main steps of the algorithm, an initial configuration was employed. Specifically, for the filtering step, Principal Component Analysis (PCA), Isomap, and Kernel Principal Component Analysis (KernelPCA) with a radial basis function kernel were selected as dimensionality reduction techniques, each producing an output dimension of 2. For the covering step, two commonly used values were chosen for the parameters defining the number of intervals (or cubes) and the percentage of overlap between them (i.e., n_cubes and perc_overlap respectively). Density-Based Spatial Clustering of Applications with Noise (DBSCAN) was utilized for the clustering step.

Table 1. Mapper algorithm parameters setting—The table presents the parameters used for the three main steps of the Mapper algorithm: filtering, covering, and clustering.

Filtering	Covering	Clustering
PCA	`n_cubes` = 20	DBSCAN with
Isomap	`perc_overlap` = 0.6	`min_samples` = 3, eps = 32
KernelPCA (kernel=rbf)		

2.2 Mapping Framework

The initial stage of our mapping framework involves constructing the graph using the Mapper algorithm. Graph construction is based exclusively on the n train data samples, ensuring that each step of the algorithm is fitted solely to this dataset. Once the graph has been created, the next step is to utilize the filter function f and cover fitted during graph construction. The framework then identifies the cover intervals that contained filtered train data points, called *hypercubes*. Nodes are formed by the samples within the same hypercube. After identifying the hypercubes, each test sample is transformed using the filter function (PCA, Isomap, or KernelPCA) and then mapped into the cover. The intervals into which each sample has been mapped are subsequently determined, allowing us to ascertain the specific node of the graph to which each test sample has been assigned. During the graph construction process in the Mapper algorithm, the cover is fitted to identify the hypercubes [1]. This involves determining the centers and the radius of the cubes (denoted as $\mathbf{C}$ and $\mathbf{r}$, respectively) in each dimension of the filtered train data space ($\mathbf{X} \in \mathbb{R}^{n \times 2}$), which, in this case,

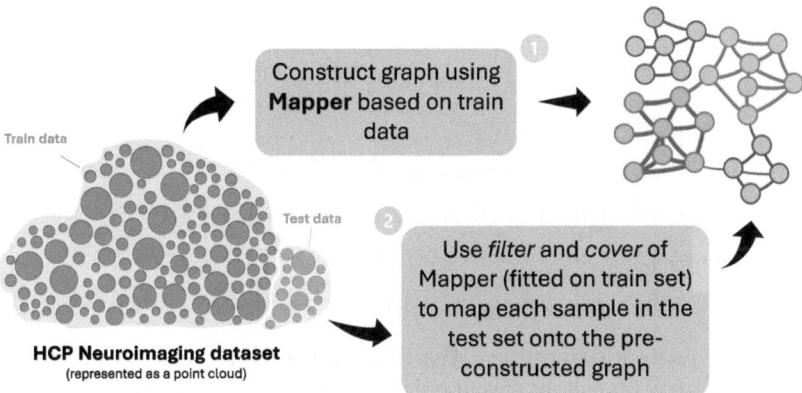

Fig. 2. Mapping framework scheme—This figure illustrates the framework scheme. It begins with the construction of a graph from the train data (1) and continues with the projection of each test sample onto the pre-constructed graph (2) to determine their node assignments.

is two-dimensional. The identification relies on the two parameters previously introduced, n_cubes ($\in \mathbb{N}$) and perc_overlap ($\in \mathbb{R}$) (Fig. 2).

As the initial step, the following two-dimensional vectors are computed, each belonging to $\mathbb{R}^{1\times 2}$:

$$\mathbf{r} = \frac{1}{2 \cdot \text{n\_cubes} \cdot (1 - \text{perc\_overlap})} \cdot \mathbf{s} \tag{1}$$

$$\mathbf{p} = \frac{\text{n\_cubes} - 1}{\text{n\_cubes}} \cdot \mathbf{s} \tag{2}$$

$$\mathbf{l} = \frac{1}{2} \cdot (\mathbf{s} - \mathbf{p}) \tag{3}$$

Here, $\mathbf{r}$ represents the radius, $\mathbf{s}$ denotes the difference between the maximum and minimum values of the filtered data in each dimension, $\mathbf{p}$ is the effective range within which the centers of the intervals are distributed and it is calculated by scaling the total data range by a factor that accounts for the number of cubes and ensures the specified overlap between adjacent cubes. The value $\mathbf{l}$ is the margin added to each side of $\mathbf{p}$ to ensure that the cubes are appropriately scaled to fit within the total filter data range.

Centers, denoted as $\mathbf{C}$, are then computed by generating a series of n_cubes evenly spaced points, separated by a distance $\mathbf{d}$, between the lower and upper bounds (α and ω, respectively) of the filtered data range in each of the two dimensions ($\mathbf{c}_1, \mathbf{c}_2$), adjusted by $\mathbf{l}$. As with (1), (2), and (3), $\mathbf{d}$, α, and ω are two-dimensional vectors defined in $\mathbb{R}^{1\times 2}$. In contrast $\mathbf{c}_1$ and $\mathbf{c}_2$ are one-dimensional vectors for each dimension of the filtered data, defining the coordinates of the centers, with a shape of $\mathbb{R}^{\text{n\_cubes}\times 1}$.

Subsequently, centers are computed by taking the Cartesian product of $\mathbf{c}_1$ and $\mathbf{c}_2$, resulting in a total of n_cubes^2 centers:

$$\begin{aligned}
\alpha &= \left\{ \min_n \{\mathbf{x}_{nk}\} + l_k \mid \mathbf{x}_{nk} \in \mathbf{X}, \ k = 1, 2 \right\} \ \in \mathbb{R}^{1\times 2} \\
\omega &= \left\{ \max_n \{\mathbf{x}_{nk}\} - l_k \mid \mathbf{x}_{nk} \in \mathbf{X}, \ k = 1, 2 \right\} \ \in \mathbb{R}^{1\times 2}
\end{aligned} \tag{4}$$

$$\mathbf{d} = \frac{1}{\text{n\_cubes} - 1} \cdot (\omega - \alpha) \ \in \mathbb{R}^{1\times 2} \tag{5}$$

$$\begin{aligned}
\mathbf{c}_1 &= \{\alpha_1 + i d_1 \mid i = 0, 1, \ldots, (\text{n\_cubes} - 1)\} \in \mathbb{R}^{\text{n\_cubes}\times 1} \\
\mathbf{c}_2 &= \{\alpha_2 + i d_2 \mid i = 0, 1, \ldots, (\text{n\_cubes} - 1)\} \in \mathbb{R}^{\text{n\_cubes}\times 1}
\end{aligned} \tag{6}$$

$$\mathbf{C} = \mathbf{c}_1 \times \mathbf{c}_2 \quad \text{with } \mathbf{C} \in \mathbb{R}^{\text{n\_cubes}^2 \times 2} \tag{7}$$

With the centers determined, the next step involves calculating the lower and upper bounds of each hypercube using the centers $\mathbf{C}$ and the radius $\mathbf{r}$. This calculation defines the spatial extent of the hypercubes in each dimension of the filtered data. Each hypercube h_j ($j = 0, 1, \ldots, \text{n\_cubes}^2 - 1$) will include the train data samples that fall within its specific bounds.

$$h_j = \{\mathbf{x}_n \in \mathbf{X} \mid \mathbf{C}_j - \mathbf{r} \leq \mathbf{x}_n \leq \mathbf{C}_j + \mathbf{r}\} \tag{8}$$

Then, when mapping m new samples from the filtered test data space ($\mathbf{X}' \in \mathbb{R}^{m \times 2}$), appropriately transformed using the filter function f fitted on train data, the centers and the radius computed during the training phase are employed to determine the specific hypercubes into which these samples will fall.

$$h_j = \left\{ \mathbf{x}_m \in \mathbf{X}' \mid \mathbf{C}_j - \mathbf{r} \leq \mathbf{x}_m \leq \mathbf{C}_j + \mathbf{r} \right\} \tag{9}$$

It is crucial to note that not all hypercubes will become nodes of the graph. Hypercubes containing a number of train samples less than the minimum number of samples parameter (`min_samples`) set for the clustering phase will be treated as noise and subsequently discarded during clustering.

After the identification of non-noisy hypercubes and the subsequent verification of single-cluster detection by DBSCAN within each of these hypercubes, we can extend the findings delineated in Eq. (9). This extension permits the assertion that these non-noisy, single-cluster hypercubes are in direct correspondence with the nodes of the previously constructed graph (Fig. 3).

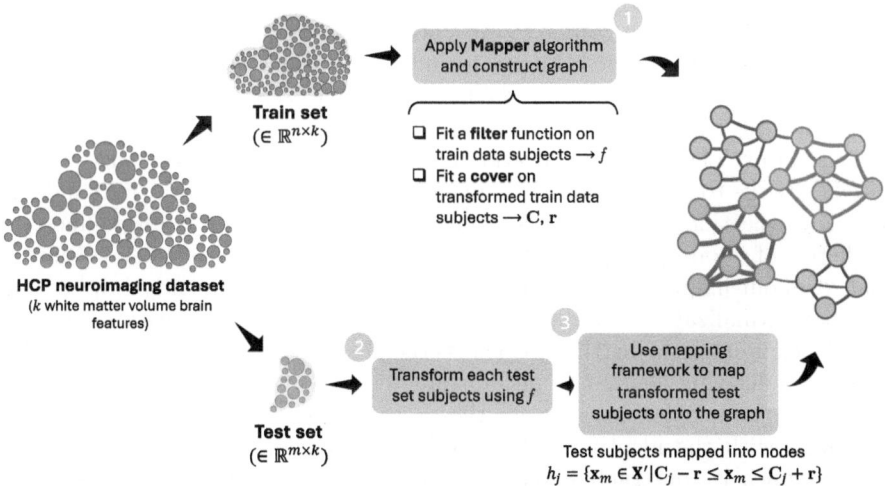

Fig. 3. Detailed scheme of the mapping framework pipeline—This figure illustrates the detailed scheme of the mapping framework pipeline. Following the division into train and test sets, 1) the Mapper algorithm is applied exclusively to the train data subjects. This process results in the construction of a graph, along with filters and covers fitted to the train data samples. Subsequently, 2) each subject in the test set is transformed using the previously fitted filter function f, and finally, 3) mapped onto the preconstructed graph by retrieving nodes through the application of Eq. (9).

Alongside the designed framework, a validation pipeline has been developed to demonstrate its reliability. Specifically, the aforementioned process has been applied to map train subjects. This procedure initially entails the construction of a graph based on the train data samples. During this phase, the filter function and the cover are fitted using the train data points, which are essential for the subsequent application of the previously delineated mapping framework.

Through the implementation of the specified methodology to the train subjects, we ascertain the nodes to which they have been allocated. The reliability of the framework can subsequently be evaluated by verifying that each subject in the train set has been mapped to the precise nodes of the graph to which they were originally assigned, as determined by the Mapper algorithm. This comparative analysis serves as a critical measure of the framework consistency and accuracy in its application to new, independent test subjects.

2.3   Graph-Based Prediction

Two applications have been developed to support the framework rationale and evaluate its reliability. We assessed the possibility to predict outcomes of interest using the Mapper graph built on regional brain features, selecting outcomes that are known to be related to the features given as input. One application involves predicting the sex of subjects within train and test sets, thereby dealing with a categorical variable. This choice of sex prediction serves as an exemplar of a binary classification task, offering a straightforward yet fundamental demographic characteristic that is known to influence various aspects of brain structure and function. Moreover, sex prediction provides a readily verifiable ground truth, facilitating the assessment of the framework performance in classification scenarios. The second application involves predicting a continuous variable, specifically the total intracranial volume (TIV) of subjects in both train and test sets. The selection of TIV serves a dual purpose. Primarily, it exemplifies the prediction of a continuous variable, thus representing a regression task within the context of our mapping framework. Additionally, TIV is frequently employed as a global normalization variable when analyzing brain volume features. Consequently, examining its distribution across the dataset may yield valuable insights into the underlying neuroanatomical characteristics of the study population.

The methodologies designed for these two applications are nearly identical, beginning with the identification of nodes *membership*. For each node of the graph—which is constructed solely based on train subjects—we identified the subjects it contains. For the first application, we determined the predominant sex for each node based on the prevalence of males and females within the node. For TIV prediction, we computed the average TIV for each node of the graph.

Starting from the predominant sex and mean TIV associated to each node, the subsequent phase involved making predictions for the train subjects. Given that each subject could be assigned to multiple nodes—a consequence inherent to the Mapper algorithm—the outcome prediction was made by considering the predictions of the relative nodes. For sex prediction, the majority voting was implemented, based on the predominant To predict the sex, we employed majority voting based on the predominant sexes associated with the nodes to which the train subjects were assigned. In contrast, TIV prediction was achieved by averaging the mean TIVs corresponding to the nodes to which the train subjects were allocated.

Subsequently, test subjects were mapped onto each graph using the proposed framework resulting in a list of nodes from the graph to which each subject was mapped. Utilizing this list of nodes, we predicted the sex and TIV of test subjects in a manner analogous to that used for the train set.

Finally, to evaluate the robustness and reliability of our proposed framework, we computed performance metrics (accuracy, f1-score, precision, and recall) for the sex prediction application, while for the TIV prediction application, we assessed the significance of the results by conducting the Wilcoxon signed-rank test on the predictions. This test examined whether the true TIV and the predicted TIV belong to the same distribution in both the train and test sets, thereby confirming the framework reliability.

3   Mapping Framework Applications

To evaluate the proposed framework, we utilized a neuroimaging dataset from the HCP, consisting of structural magnetic resonance imaging (sMRI) data and demographic variables from 1109 healthy subjects (505 males, 604 females). The sMRI scans were acquired using T1-weighted sequences on 3T MRI scanners. Preprocessing followed a Voxel-Based Morphometry (VBM) protocol using the Computational Anatomy Toolbox (CAT12) [2] within SPM12 [6], including tissue segmentation, spatial normalization, modulation, and spatial smoothing. White matter morphological features were extracted, identifying 52 white matter volume (WMV) measures from the CoBra subcortical regional atlas [11].

For each of the two applications, the dataset was divided into train and test sets in an 80% to 20% ratio stratified by sex, respectively. Following the dataset division in both applications, a graph was constructed using the Mapper algorithm, based on train data samples only and the parameters detailed in Table 1. Specifically, three distinct graphs were generated, each corresponding to one of the three dimensionality reduction techniques employed in the filtering step of the Mapper algorithm (PCA, Isomap, and KernelPCA), and depicted using a force-directed layout within the DyNeuSR platform, with each node represented by pie charts (see Fig. 4 and 5). The validation pipeline described in Sect. 2.2 has been applied to all three graphs. Each subject in the train set used to construct the graphs has been accurately mapped to the correct nodes, thereby demonstrating the reliability of the framework.

4   Results Discussion

In both applications, the findings demonstrate the reliability of our framework. For sex prediction, similar classification performance is observed in both train and test sets, indicating non-overfitted results (see Table 2). In the second application, the Wilcoxon test (see Table 3) showed non-significant *p-values*, as expected. Together, these two applications demonstrate the framework versatility in handling both categorical and continuous variables in neuroimaging contexts.

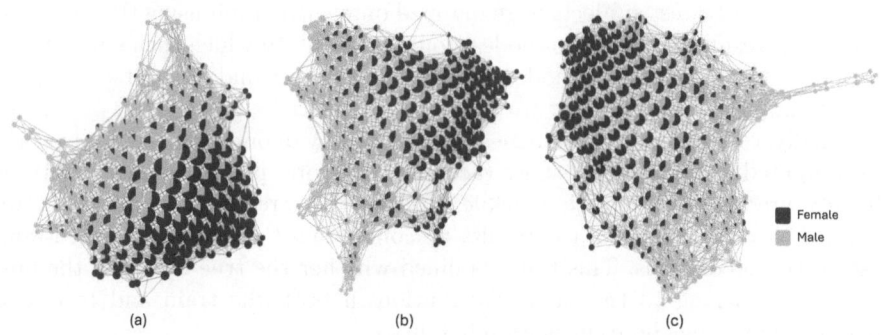

(a)   (b)   (c)

Fig. 4. Graphical depiction of sex distribution within train set—This figure presents three distinct graphs constructed from train data samples using different dimensionality reduction techniques in the filter step of the Mapper algorithm: (a) PCA, (b) Isomap, and (c) KernelPCA. In all three cases, the covering and clustering parameters are those specified in Table 1. Nodes are color-coded to illustrate the distribution of sex across the train set.

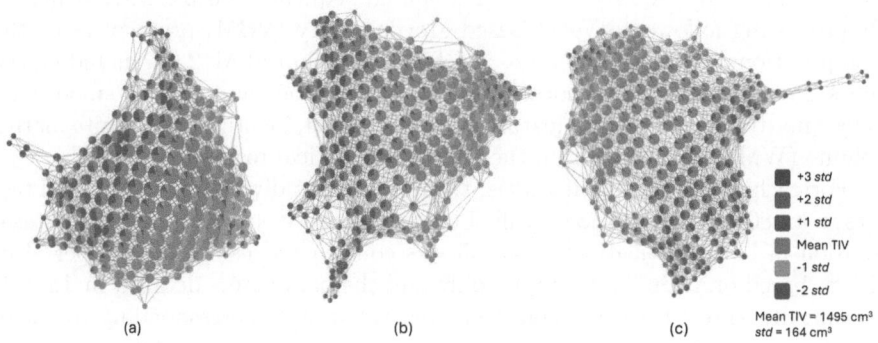

(a)   (b)   (c)

Fig. 5. Graphical depiction of TIV variation within train set—This figure presents three distinct graphs constructed from train data samples using different dimensionality reduction techniques in the filter of the Mapper algorithm: (a) PCA, (b) Isomap, and (c) KernelPCA. In all three cases, the covering and clustering parameters are those specified in Table 1. Nodes are color-coded to illustrate the distribution of TIV across the train set.

Table 2. Graph-based sex prediction results—The table presents the classification performance of various graphs constructed using different filter functions (train | test)

Filter function	Accuracy	f1-score*	Precision*	Recall*
PCA	0.795 | 0.743	0.793 | 0.742	0.798 | 0.743	0.795 | 0.743
Isomap	0.799 | 0.739	0.796 | 0.736	0.804 | 0.740	0.799 | 0.739
KernelPCA	0.794 | 0.739	0.792 | 0.737	0.798 | 0.739	0.794 | 0.739

*weighted

Table 3. Graph-based TIV prediction results—This table presents the *p-values* from the Wilcoxon test, which was conducted to determine whether the true and predicted TIVs originate from the same distribution (train | test).

Filter function	*p-value*
PCA	0.508 | 0.778
Isomap	0.909 | 0.872
KernelPCA	0.586 | 0.789

It is important, however, to emphasize that the main focus should not be on application performance scores but rather on the reliability of the framework, demonstrated by the similarity of results in train and test sets. This novel framework within TDA represents a significant innovation, enabling the integration of new data samples onto a pre-constructed graph for sophisticated analysis of complex datasets. This approach yields profound insights into the relationships, similarities, and disparities between newly introduced data and established topological features. The aim is to extend the framework to various neuroimaging applications, to extract clinically relevant features that enable more accurate and effective predictions.

5   Conclusions

In this study, we have demonstrated the effectiveness and reliability of our novel framework for neuroimaging analysis using TDA. Our approach leverages the Mapper algorithm to construct graphs from train data, onto which new independent samples of a test set are mapped to predict outcomes. Notably, we extended the Mapper tool to transfer unknown test data to a pre-existing graph, paving the way for the use of Mapper in cross-validation frameworks.

Applying our framework to sex and TIV prediction in HCP neuroimaging dataset yielded promising results. Consistent performance metrics between train and test sets for sex prediction and non-significant *p-values* from the Wilcoxon test for TIV prediction validate the framework robustness. Emphasizing reliability and consistency over performance scores, the framework shows potential for diverse neuroimaging applications and extracting clinically relevant features for more accurate predictions.

Code available at: github.com/brainpolislab/TDA-Graph-Mapping-Framework

Acknowledgments. Data were provided by the Human Connectome Project, WU-Minn Consortium (Principal Investigators: David Van Essen and Kamil Ugurbil; 1U54MH091657) funded by the 16 NIH Institutes and Centers that support the NIH Blueprint for Neuroscience Research; and by the McDonnell Center for Systems Neuroscience at Washington University.

The study was partially supported by the Italian Ministry of University and Research (grant n. 2022RXM3H7 to EM, P20229MFRC to SV) and by the Italian Ministry of Health (grant n. GR-2019-12370616 to EM).

This study was partially supported by the EBRAINS-Italy, project funded under the National Recovery and Resilience Plan (NRRP), Mission 4, "Education and Research" - Component 2, "From research to Business" Investiment 3.1 - Call for tender No. 3264 of Dec 28, 2021 of Italian Ministry of University and Research funded by the European Union - NextGenerationEU, with award number: Project code IR0000011, Concession Decree No. 117 of June 21,2022 adopted by the Italian Ministry of University and Research, CUP B51E22000150006, Project title "EBRAINS-Italy" (European Brain ReseArch INfrastructureS-Italy).

References

1. Carrière, M., Michel, B., Oudot, S.: Statistical analysis and parameter selection for mapper. J. Mach. Learn. Res. **19**(12), 1–39 (2018). http://jmlr.org/papers/v19/17-291.html

2. Gaser, C., Dahnke, R., Thompson, P.M., Kurth, F., Luders, E., Initiative, A.D.N.: CAT–a computational anatomy toolbox for the analysis of structural MRI data. biorxiv pp. 2022–06 (2022). https://doi.org/10.1101/2022.06.11.495736

3. Geniesse, C., Sporns, O., Petri, G., Saggar, M.: Generating dynamical neuroimaging spatiotemporal representations (DyNeuSR) using topological data analysis. Netw. Neurosci. **3**(3), 763–778 (2019). https://doi.org/10.1162/netn_a_00093

4. Gidea, M., Katz, Y.: Topological data analysis of financial time series: landscapes of crashes. Physica A Stat. Mech. Appl. **491**, 820–834 (2018). https://www.sciencedirect.com/science/article/pii/S0378437117309202

5. Nielson, J.L., et al.: Topological data analysis for discovery in preclinical spinal cord injury and traumatic brain injury. Nat. Commun. **6**(1), 8581 (2015). https://doi.org/10.1038/ncomms9581

6. Penny, W.D., Friston, K.J., Ashburner, J.T., Kiebel, S.J., Nichols, T.E.: Statistical Parametric Mapping: The Analysis of Functional Brain Images. Elsevier, Amsterdam (2011)

7. Saggar, M., Shine, J.M., Liégeois, R., Dosenbach, N.U., Fair, D.: Precision dynamical mapping using topological data analysis reveals a hub-like transition state at rest. Nat. Commun. **13**(1), 4791 (2022). https://doi.org/10.1038/s41467-022-32381-2

8. Singh, G., Mémoli, F., Carlsson, G.E., et al.: Topological methods for the analysis of high dimensional data sets and 3D object recognition. PBG@ Eurograph. **2**, 091–100 (2007)

9. Skaf, Y., Laubenbacher, R.: Topological data analysis in biomedicine: a review. J. Biomed. Inform. **130**, 104082 (2022). https://doi.org/10.1016/j.jbi.2022.104082

10. Smith, A.D., Dłotko, P., Zavala, V.M.: Topological data analysis: concepts, computation, and applications in chemical engineering. Comput. Chem. Eng. **146**, 107202 (2021). https://doi.org/10.1016/j.compchemeng.2020.107202

11. Tullo, S., Devenyi, G.A., Patel, R., Park, M.T.M., Collins, D.L., Chakravarty, M.M.: Warping an atlas derived from serial histology to 5 high-resolution MRIs. Sci. Data **5**(1), 1–10 (2018). https://doi.org/10.1038/sdata.2018.107

12. Van Essen, D.C., et al.: The WU-Minn human connectome project: an overview. Neuroimage **80**, 62–79 (2013). https://doi.org/10.1016/j.neuroimage.2013.05.041

13. van Veen, H.J., Saul, N., Eargle, D., Mangham, S.W.: Kepler mapper: a flexible python implementation of the mapper algorithm. J. Open Source Softw. **4**(42), 1315 (2019). https://doi.org/10.21105/joss.01315

14. Wasserman, L.: Topological data analysis. Ann. Rev. Stat. Appl. **5**, 501–532 (2018). https://doi.org/10.1146/annurev-statistics-031017-100045
15. Yao, Y., et al.: Topological methods for exploring low-density states in biomolecular folding pathways. J. Chem. Phys. **130**(14) (2009). https://doi.org/10.1063/1.3103496

Topological Data Analysis of Resting-State fMRI Suggests Altered Brain Network Topology in Functional Dyspepsia: A Mapper-Based Parcellation Approach

Emma Tassi[1,2,3](✉) ⓘ, Harrison Fisher[4], Andrew Bolender[5], Jun-Hwan Lee[6], Lizbeth J. Ayoub[2] ⓘ, Anna Maria Bianchi[3] ⓘ, Braden Kuo[5] ⓘ, Vitaly Napadow[2,4] ⓘ, Eleonora Maggioni[3] ⓘ, and Roberta Sclocco[2,4,5] ⓘ

[1] Department of Neuroscience and Mental Health, Fondazione IRCCS Ca' Granda, Osp. Maggiore Policlinico, Milan, Italy
emma.tassi@polimi.it
[2] Spaulding Rehabilitation Hospital, Harvard Medical School, Boston, MA, USA
[3] Department of Electronics, Information and Bioengineering, Politecnico di Milano, Milano, Italy
[4] Athinoula A. Martinos Center for Biomedical Imaging, Massachusetts General Hospital, Harvard Medical School, Boston, MA, USA
[5] Department of Medicine, GI Unit, Massachusetts General Hospital, Harvard Medical School, Boston, MA, USA
[6] Korea Institute of Oriental Medicine, Daejeon, Korea

Abstract. Functional dyspepsia (FD) is a complex condition identified by chronic indigestion without an obvious organic cause, characterized by diverse abdominal symptoms. Recent studies employing resting-state functional magnetic resonance imaging (rs-fMRI) have investigated gut-brain interactions in FD. These studies report altered functional connectivity patterns that are associated with the severity of the disease. The investigation of resting-state functional connectivity patterns involves defining connectivity nodes for subsequent graph-theory analyses, thus emphasizing the importance of brain parcellation. While traditional methods employ predefined brain atlases, fMRI-driven parcellation offers a more specific approach able to extract functionally homogeneous regions. In this study, we applied the Topological Data Analysis (TDA) tool of Mapper algorithm to rs-fMRI data to develop a whole-brain TDA-driven fMRI parcellation pipeline. This functional parcellation, applied in a group of healthy controls (HC), provides a reference for comparing network properties between HC and FD groups. We propose that the TDA Mapper is able to recover structure in rs-fMRI data, showing that topological complexes embedded in fMRI data could be identified and explored using this tool. Based on the brain network thus derived, we highlight the potential of applying graph analysis on rs-fMRI data to assess topological properties of brain connectivity, showing significant differences between groups in the functional parcel located in the frontal pole for nodal strength and degree.

Keywords: Topological data analysis · Functional dyspepsia · Resting-state fMRI

1 Introduction

Recent advances in neuroimaging, especially resting-state functional magnetic resonance imaging (rs-fMRI), have provided crucial insights into brain-gut interactions in Functional Dyspepsia (FD), a complex and symptom-based disorders characterized by dysfunction along the brain-gut axis [1]. Rs-fMRI measures spontaneous brain activity by detecting fluctuations in blood oxygen level-dependent (BOLD) signals, revealing the brain's intrinsic organization and functional connectivity at rest. Selective connectivity alterations in brain regions such as the thalamus, somatosensory cortex, prefrontal cortex, insula, and anterior cingulate cortex suggest that FD may involve dysregulation in networks modulating visceral pain and discomfort [1]. Investigating resting-state functional connectivity patterns involves defining nodes for connectivity and graph-theory analyses, emphasizing the importance of brain parcellation. Functional connectivity and network analysis typically uses predefined brain atlases like Automated Anatomical Labeling Atlas (AAL) or Broadman to provide network's nodes, though these may not fully capture the brain's functional organization, highlighting the need for fMRI-driven brain parcellation [2]. On the other hand, functional parcellation techniques divide the brain into regions with similar BOLD signals, and provide novel insight into brain organization and diverse brain disorders [2,3]. These functional parcellation atlases are valuable for comparing network properties between healthy controls (HC) and patient groups [4]. The applications of fMRI-driven parcellations span from whole-brain methods to specific region of interest parcellations, including seed-guided parcellations that use a seed region to initialize functional parcels [5]. Whole-brain parcellations provide comprehensive connectivity maps but are computationally demanding, while seed-guided parcellations balance detailed analysis and computational efficiency. Both functional and anatomical atlases have been used for identifying seed regions, though the optimal parcellation method remains under debate. Performance assessments of different parcellation strategies, often involving regional homogeneity estimates, are essential. Numerous data-driven methods for functional parcellation, such as k-means clustering, spectral clustering [3], Gaussian mixture models, independent component analysis (ICA), Markov random fields [6], and topological data analysis (TDA) [7], have been used to create functional atlases at both individual and group levels. Among these methodologies, TDA has emerged as a powerful tool for fMRI data analysis, due to its capability of handling high-dimensional, nonlinear datasets, capturing patterns across multiple scales and maintaining stability under perturbations [8]. TDA comprises a set of methods grounded in the algebraic topology field, defined as the area of mathematics that focuses on the notion of shape and connectivity, and attempts to study patterns (i.e., features) within the data. Specifically, TDA studies the "shape" (i.e., underlying topology) of the data and is optimized to search for latent structures, also called topological structures. TDA methods of persistent homology and Mapper effectively reveal dynamic structures in fMRI data. TDA Mapper, introduced by Singh et al. [9], is one of the most commonly used topological data analysis tools and aims to identify topological structure within large-scale datasets, allowing direct data visualization and exploration. Indeed, TDA has been shown to effectively identify clusters of functionally related subregions [8], employing both persistent homology [7] and Mapper tool [10]. Specifically, the TDA Mapper visualizes complex data shapes and extracts topologically relevant features. These properties motivate further exploration of this tool in retrieving relevant topological structure embedded in fMRI data. In this work, we applied the TDA Mapper algorithm to rs-fMRI data to develop a whole-brain TDA-driven fMRI parcellation pipeline. This functional parcellation, applied to rs-fMRI data from HC, serves as a reference framework for comparing network properties between HC and FD groups, offering new insights into FD neural underpinnings. Our approach is a seed-guided parcellation generating functional parcels based on both anatomical and functional information, while respecting anatomical boundaries to avoid meaningless parcels. Parcels were constrained within Harvard-Oxford (HO) anatomical regions, and voxels with homogenous functional patterns were grouped together [11]. The performance of the extracted parcellation was assessed based on two cluster validity indexes and compared with two standard atlases. Further, graph-theory analysis was applied on FD and HC connectivity matrices. Global and local topological characteristics that were significantly different between the two groups were extracted. The results demonstrated TDA Mapper's potential in identifying functionally segregated subregions from rs-fMRI, and applicability in examining network heterogeneity in clinical populations. We also identified different topological properties between FD and HC groups, thus adding insights to FD characterization by exploiting multi-scale topological approach.

2   Methods

2.1   Participants and Study Protocol

The dataset considered in this study includes 10 HC and 10 FD patients aged 21–65 years (HC: age mean $\pm$ std, 32.80 $\pm$ 6.30; M/F, 3/7 and FD: 32.80 $\pm$ 14.85, M/F, 1/9). Inclusion and exclusion criteria for FD and HC patients, as well as study protocol details, were previously described [12]. Briefly, both groups underwent three serial gastric MRI and brain fMRI scans following ingestion of a food-based contrast meal. For this study, brain fMRI data acquired from the first scan immediately following meal ingestion was considered.

2.2   Brain fMRI Data Acquisition and Processing

Blood-oxygenation level-dependent (BOLD) fMRI data were collected on a Siemens Skyra 3T MRI scanner, using a sixty-four-channel head/neck coil. Gradient-echo echo-planar imaging and multi-slice acquisition with multiband factor 5 were used to collect whole-brain fMRI data ($2.04 \times 2.04 \times 2.00$ mm voxel size, 75 axial slices, repetition time 1270 ms, echo time 33 ms, flip angle 65°, 288 timeseries measurements). Further details related to MRI and fMRI acquisition parameters were previously described [12], as well as a full description of fMRI data preprocessing steps.

2.3   Whole-Brain TDA-Driven fMRI Parcellation Based on Control Group

Group-Wise Brain Functional Parcellation Based On TDA Mapper. Specific details of TDA Mapper's pipeline and its application on fMRI data were previously described [9]. Briefly, TDA Mapper algorithm started by applying a filter function to map high-dimensional data into a lower dimensional space. The filtered data were then divided into overlapping intervals defining a cover, that was characterized by the parameters of Resolution (R, number of overlapping bins) and Gain (G, overlap between bins). Clustering was then applied to raw data within each overlapping bin. Each cluster of data formed a node of the graph, and edges were traced among overlapping clusters. In our study, TDA Mapper tool provided by Kepler library [13] was applied, as well as NetworkX library [14] for extracting community structures within the graph. Specifically, community detection (CD) analysis relies on the idea that networks usually show a hierarchical and modular organization [15]. The network's community structures capture relevant features and are often considered as subgraphs of the original network, characterized by nodes that are densely connected. CD analysis has been applied in diverse fields, as biochemical or neural networks, to retrieve communities that might be regarded as unique functional groups within the network. We implemented a whole-brain TDA-driven fMRI parcellation to retrieve a functional parcellation representative of a reference control population. Our parcellation pipeline was composed of 3 main steps: data masking, 1st level, and 2nd level TDA Mapper application.

Anatomical Masking. First, the whole brain fMRI volume was anatomically parceled into 48 regions based on the HO cortical atlas [16]. This approach prevented the identification of irrelevant parcels, such as those that encompass multiple anatomically distinct regions [17].

1st level TDA Mapper. The preprocessed 4D fMRI data within each HO region were transformed in a 2D matrix, with rows representing voxels in the specific HO region, and columns representing time frames (TR) for all concatenated HC subjects. This matrix was fed into the 1st level TDA Mapper application (TDA Mapper's input dimension: # rows: #Voxels of HO region considered; #columns:

$\#TR \times \#HCsubjects$; where $\times$ refers to the product). For each HO region, the resulting TDA Mapper graph created nodes grouping voxels characterized by similar activity patterns, and edges were created between two nodes sharing at least one voxel. Then, the CD algorithm was applied to extract relevant groups of nodes within the graph (i.e., groups of clusters of voxels). Communities in the graph captured relevant topological features and were composed by graph's nodes densely connected. Indeed, these communities, reflecting the hierarchical organization of the network, are composed of densely connected nodes sharing common properties and/or playing similar roles within the network [15]. CD algorithm of Label Propagation [18] was applied to extract community structures within the graph by maximizing modularity. As a result, we retrieved the groups of graph's nodes composing each community structure identified. Thus, the clusters of voxels contained in the nodes, included in each community, were grouped together. Each group represented a single *subcluster* within the specific HO region. Indeed, a *subcluster* coincided with a single community and indicated a smaller group of voxels belonging to the cluster of HO region. The voxels composing each community were characterized by a densely connected topological structure and coherent temporal profiles (i.e., functional activity).

2nd level TDA Mapper. The 2nd TDA Mapper received as input the 2D matrix containing the mean voxels' functional activity in the subclusters found as output of 1st level TDA Mapper, where rows indicate all the subclusters and columns represent TR collected for all HC subjects (TDA Mapper's input dimension: #rows: #Subclusters found in all HO regions; #columns: $\#TR \times \#HCsubjects$). The hub-node (i.e., a single *group of subclusters*) of the resulting graph, indicative of the most influential node, was then extracted as the node associated with the highest average degree and closeness centrality measures. This specific node, composed by a group of subclusters of voxels, represented the final set of functional regions outlying the reference-based atlas. These functional regions were characterized by coherent functional activity and were encapsulating key topological properties of the graph. The schematic pipeline of the parcellation algorithm is reported as follows.

Detailed TDA Mapper's Parameters. Euclidean distance was chosen as the metric of distance (i.e., similarity) between data points. Input group-data matrix was standardized and the original patterns were projected along two dimensions of the neighborhood lens function (tSNE). Filter's ranges were divided into overlapping bins by defining R and G parameters, controlling respectively for the number of bins and overlap among them. Cover parameters were selected by testing a wide range of R-G pairs on the input matrix and selecting based on visual inspection of the most stable graph construction. Finally, density-based spatial clustering (DBSCAN) was performed to cluster subjects in the same bin and trace an edge between pairs of overlapping clusters (i.e., that share voxels). The unique cover parameter solution R = 4 and G = 0.85 was selected for 1st level TDA Mapper for each H-O region, whereas R-G pairs of R = 9 and G = 0.80 were applied for the 2nd level TDA Mapper.

Algorithm 1. Whole-brain TDA-driven parcellation

---

procedure FUNCTIONAL PARCELLATION WITHIN HARVARD-OXFORD (HO) REGIONS

    Input: 4D BOLD time series of HC group

    `for i=1:N regions of HO do`   ▷ Anatomical Masking

        `Mask 4D BOLD time series in i region`

    `Concatenate Masked 4D BOLD time series of all HC`

    t = Group-level Masked BOLD time series matrix for all HO regions

    `for i=1:N regions of HO do`

        `Computation TDA Mapper graph on t(i)`   ▷ 1st level TDA

        `Application of community detection algorithm`

        `Identification of communities`

        `Identification of subclusters' voxels within communities' nodes`

        `Computation of mean activation within subclusters found`

    k = Group-level mean activation matrix within subclusters

    `Computation of TDA Mapper graph on k`   ▷ 2nd level TDA

    `Identification of the hub-node`

    `Identification of subclusters of voxels within hub-node`

    Output: Group-wise parcellation

Characteristics of functional parcels extracted. The proposed parcellation framework could result, for each HO region, in (i) one or more functional parcels, or (ii) no functional parcels. Indeed, the purpose of the 2nd level TDA Mapper was to topologically extract a set of subclusters among the ones found in all HO regions, without the constraint of the existence of at least one parcel in each HO region. Further, no spatial constraints were applied during the TDA Mapper clustering step, thus resulting in clusters entirely based on functional profiles information [3].

Performance Assessment. Parcellation reliability, in terms of homogeneity and region separation criteria, was assessed by cluster reproducibility indices of Kendall's coefficient of concordance (KCC) [19] and Silhouette index (SI) extracted for each parcel [20]. These same indices extracted were also calculated on parcellations from the HO and AAL atlases for comparison. The homogeneity estimate with KCC was calculated between the BOLD time series of all voxels within a parcel, obtaining one value for each parcel, and then averaged across parcels, resulting in a single value. SI was computed for each parcel by comparing within-parcel dissimilarity and inter-parcel dissimilarity across all voxels [2], thus obtaining a measure of compactness of the parcel and degree of separation among parcels. The distance measure defining dissimilarity was based on the Pearson correlation coefficient calculated between each voxel time series. For the three parcellations, within-, inter-parcel dissimilarity, and SI were calculated for each parcel, as well as an average SI value among clusters. Nonparametric Kruskal Wallis test was applied to test for statistical difference in KCC and SI distributions among parcels of the three parcellations. As follows, if significant differences emerged ($p < .05$, Bonferroni corrected with $n = 3$, number of parcellation compared), post-hoc multiple pairwise comparison results were performed.

2.4 ROI-to-ROI Functional Connectivity Analysis

ROI-to-ROI functional connectivity analyses were applied using SPM Marsbar toolbox on both HC and FD groups employing the functional parcels obtained on the reference HC group. For each subject, statistical dependencies among ROIs were assessed using Pearson correlation coefficients, resulting in a N × N FC adjacency matrix for each participant, the elements of which represent the pairwise cross-correlation between the BOLD time series of the corresponding ROIs.

2.5 Graph Theoretical Analysis Between HC and FD Group

For each single-subject FC matrix, only positive Pearson correlation values (i.e., positive-value edges) across all elements of the matrix were selected [26]. Proportional thresholding (PTh) was applied to threshold the FC matrices by preserving a pro-portion of the strongest connections (PT%) within the PTh range of 0.06–0.3, with a step size of 0.01. This resulted in twenty-five PTh binary adjacency matrices with a density of PT% composed by a subset of links. After PTh, global and nodal topological properties of the binary adjacency matrices were extracted for the subsequent comparison between FD and HC groups using the Brain Connectivity Toolbox (BCT). Three global properties - namely, global density, global degree, and global efficiency - were extracted, as well as five nodal measures (nodal degree, nodal strength, nodal modularity, nodal efficiency and clustering coefficient). For each PTh binary adjacency matrix, one value was obtained for each global measure, and N values were extracted for each local measure. Statistical differences between FD and HC groups in global and nodal topological features were assessed by employing Wilcoxon rank non-parametric test [21]. For each PTh binary adjacency matrix, the differences of global features between groups were computed and assessed for statistical significance. Further, the nodes with significantly different nodal measures between groups were identified for each PTh matrix. To summarize statistical results across a wide range of PTh matrices, the global features that resulted significantly different among groups for more than half of the PTh matrices were selected. The node that results significantly different for most of the nodal features and over more than half of the PTh matrices was calculated and considered informative for the differentiation among groups.

3   Results

3.1 TDA-Driven fMRI Parcellation Based on Reference Group

In this group-wise functional parcellation, the reference population composed by HC subjects at rest were considered for extracting the functional parcels in the time domain, resulting in 37 distinct parcels in total. The functional parcels were identified in 32 unique regions of HO and most of the voxels assigned to each of these clusters were spatially adjacent to one another. The identified group-wise functional parcellation is shown in Fig. 1.

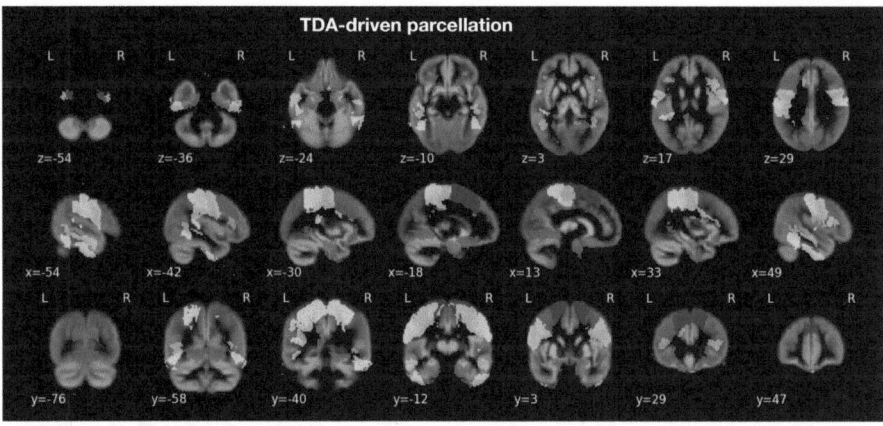

Fig. 1. Functional parcels derived from the whole-brain TDA-driven fMRI parcellation algorithm.

3.2   Performance Assessment

The resulting functional parcels were evaluated in terms of KCC and SI and compared among HO and AAL parcellations. Clustering reproducibility was extracted for each cluster and then averaged across ROIs. Our whole-brain TDA-driven fMRI parcellation exhibited higher regional homogeneity across ROIs (KCC = 0.28 ± 0.18) compared to both functional and anatomical atlas (KCC AAL = 0.21 ± 0.07, KCC HO: 0.17 ± 0.08), with a significantly higher KCC compared to HO (p < .001, Bonferroni-corrected), demonstrating high intra-cluster homogeneity among voxels across parcels. Parcel size had a strong effect on KCC (r = 0.40, p < .05), that is, the larger the cluster, the lower the regional homogeneity. As for SI, our whole-brain TDA-driven fMRI parcellation (SI = 0.25 $\pm$ 0.01) significantly outperformed both AAL (SI = 0.12 $\pm$ 0.005) and HO (SI = 0.11 $\pm$ 0.06) parcellations (p < .001, Bonferroni-corrected), indicating superior performance of the proposed approach in terms of cluster compactness and separation.

3.3   Graph Theoretical Analysis Between HC and FD Group

No global metrics were found to be significantly different among HC and FD groups for more than half of the PTh matrices. However, nodal degree and nodal strength were significantly higher in the parcel localized within the frontal pole for the FD group in more than half of the adjacency matrices (nodal degree: 96% of PTh matrices, nodal strength: 100% of PTh matrices) compared to HC. The distribution of these nodal parameters within the Frontal Pole parcel averaged across PTh matrices are shown in Fig. 2.

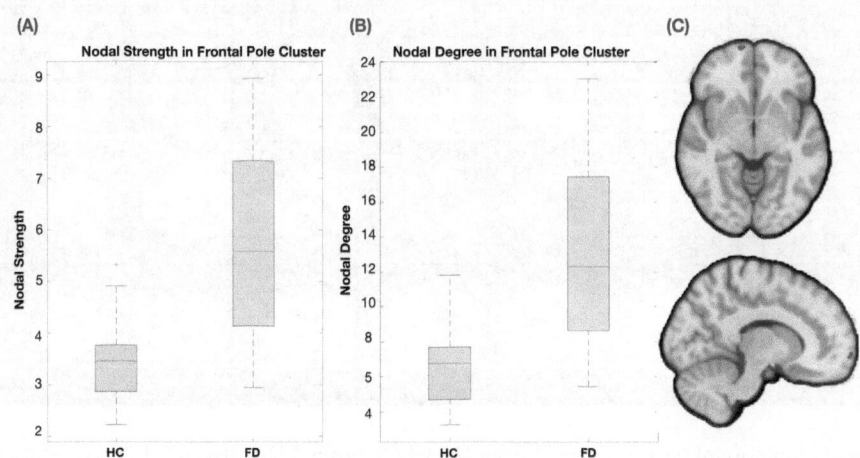

Fig. 2. Group difference in nodal measures in frontal pole parcel. Nodal strength (A) and nodal degree (B) in Frontal Pole parcel. Panel (C) shows in red the Frontal Pole parcel found by the parcellation approach and in yellow the Frontal Pole region of HO. (Color figure online)

3.4 Discussion

This study proposed a novel framework of whole-brain group-wise functional parcellation based on TDA Mapper employing rs-fMRI data, followed by a graph-based analysis to identify changes in topological characteristics among control and FD groups. The development of functional parcellation techniques is important in fMRI connectivity research to capture the brain's intrinsic functional organization. As shown here, partitioning the brain into functionally homogeneous subregions is useful to extract multivariate spatial features that reflect functional relationships among diverse brain areas. Within this context, the identification of parcels sharing functional characteristics across a healthy population is needed as a framework reference for detecting pathological alterations [4]. Our method was applied to extract functional subunits from the control group, obtaining a parcellation that serves as a reference spatial map to contrast network properties between HC and FD groups. At the first level of our pipeline, the functional clusters were delineated within each HO, and then only clusters of voxels associated with relevant community structure in the whole-brain configuration were extracted within the boundaries of a single HO region.

We avoided combining voxels from distinct anatomical regions into a single parcel, considering the structural heterogeneity of distinct anatomical regions [17]. Within the second level, only functional clusters representing the focal point of convergence/divergence of information (i.e., hub-node) in the 1st level functional clusters configuration were extracted.

The proposed whole-brain TDA-driven fMRI parcellation method was data-driven within each HO seed, since (i) it was aimed at grouping the brain into functionally relevant regions based only on fMRI times series, (ii) no minimum number of clusters was imposed and (iii) no spatial constraints within each HO seed were introduced. These conditions allowed the clustering of the voxels to rely only on functional synchrony information, rather than on spatial proximity. The proposed tool allows to capture high dimensional multivariate patterns within the brain, handling non-linearity and multimodality within the fMRI data. Of note, TDA Mapper allows for direct data exploration and extraction of topological relevant properties embedded in structured fMRI signals. Functional homogeneity within parcels, as well as parcels separation, resulted in values of clusters reproducibility indexes consistent with previous applications [2]. Moreover, this study demonstrated the direct applicability of the proposed functional parcellation within a subsequent graph-theory analysis, aimed to assess alterations in topological properties among controls and patient groups. Interestingly, we showed a significantly higher nodal degree and nodal strength in a functional parcel located in the frontal pole in FD patients, which could be interpreted as a significantly altered involvement and information flow related to this node. These results were consistent with previous findings showing aberrant processing involving frontal cortex regions, implicated in executive and integrative control functions, in patients suffering from FD [12]. In future studies, TDA-driven fMRI parcellation should be applied to larger samples and be compared with standard functional parcellation's techniques. Therefore, despite underlying the potential use of this data-driven parcellation in population-based studies of functional network, future researches should focus on its applicability for characterizing the associations among FC patterns and patients' clinical outcome.

3.5   Conclusions

This study presents a novel approach to whole-brain group-wise functional parcellation based on the technique of TDA Mapper. The proposed framework is data-driven within each HO seed, which can facilitate subsequent analyses to better investigate brain functional organization by employing multivariate techniques. We demonstrate that the TDA Mapper can successfully recover topological structures in fMRI data. Each functional parcel grouped voxels with functionally homogeneous profile and lying within one anatomical region, without spatially constraining on voxels location within each region. The proposed approach can be used as a preprocessing step for node-level identification, as well as a method for exploring and visualizing multivariate functional relationships or spatial features within rs-fMRI data.

Code available at: https://osf.io/5hwck/

Acknowledgments. The present work was supported by the following organizations: US National Institutes of Health (NIH), National Institute of Diabetes and Digestive and Kidney Diseases (U01-DK112193, R01-DK133520, R01-DK136243); NIH National Center for Complementary and Integrative Health (P01-AT009965, R21-AT011918, K01-AT012208); Osher Center for Integrative Medicine (Pilot Research Grant). EM was partly supported by the Italian Ministry of Health (grant n. GR-2019-12370616) and by the Italian Ministry of University and Research (PRIN 2022 PNRR, grant n. P20229MFRC).

References

1. Tack, J., et al.: Functional gastroduodenal disorders. Gastroenterology **130**(5), 1466–1479 (2006). https://linkinghub.elsevier.com/retrieve/pii/S0016508506005087

2. Craddock, R.C., James, G., Holtzheimer, P.E., Hu, X.P., Mayberg, H.S.: A whole brain fMRI atlas generated via spatially constrained spectral clustering. Hum. Brain Mapp. **33**(8), 1914–1928 (2012). https://onlinelibrary.wiley.com/doi/10.1002/hbm.21333

3. Beckmann, M., Johansen-Berg, H., Rushworth, M.F.S.: Connectivity-based parcellation of human cingulate cortex and its relation to functional specialization. J. Neurosci. **29**(4), 1175–1190 (2009). https://www.jneurosci.org/lookup/doi/10.1523/JNEUROSCI.3328-08.2009

4. Shen, X., Papademetris, X., Constable, R.: Graph-theory based parcellation of functional subunits in the brain from resting-state fMRI data. NeuroImage **50**(3), 1027–1035 (2010). https://linkinghub.elsevier.com/retrieve/pii/S105381190901427X

5. Iraji, A., et al.: The connectivity domain: analyzing resting state fMRI data using feature-based data-driven and model-based methods. NeuroImage **134**, 494–507 (2016). https://linkinghub.elsevier.com/retrieve/pii/S1053811916300398

6. Ryali, S., Chen, T., Supekar, K., Menon, V.: A parcellation scheme based on von Mises-Fisher distributions and Markov random fields for segmenting brain regions using resting-state fMRI. NeuroImage **65**, 83–96 (2013). https://linkinghub.elsevier.com/retrieve/pii/S1053811912009858

7. Ellis, C.T., Lesnick, M., Henselman-Petrusek, G., Keller, B., Cohen, J.D.: Feasibility of topological data analysis for event-related fMRI. Netw. Neurosci. **3**(3), 695–706 (2019). https://direct.mit.edu/netn/article/3/3/695-706/2174

8. Salch, A., Regalski, A., Abdallah, H., Suryadevara, R., Catanzaro, M.J., Diwadkar, V.A.: From mathematics to medicine: a practical primer on topological data analysis (TDA) and the development of related analytic tools for the functional discovery of latent structure in fMRI data. PLOS ONE **16**(8), e0255859 (2021). https://dx.plos.org/10.1371/journal.pone.0255859

9. Singh, G., Memoli, F., Carlsson, G.: Topological methods for the analysis of high dimensional data sets and 3D object recognition (2007). Artwork Size: 10 pages ISBN: 9783905673517 ISSN: 1811-7813 Pages: 10 pages Publication Title: Eurographics Symposium on Point-Based Graphics. http://diglib.eg.org/handle/10.2312/SPBG.SPBG07.091-100

10. Saggar, M., et al.: Towards a new approach to reveal dynamical organization of the brain using topological data analysis. Nat. Commun. **9**(1), 1399 (2018). https://www.nature.com/articles/s41467-018-03664-4

11. Rubinov, M., Sporns, O.: Complex network measures of brain connectivity: uses and interpretations. NeuroImage **52**(3), 1059–1069 (2010). https://linkinghub. elsevier.com/retrieve/pii/S105381190901074X
12. Sclocco, R., et al.: Cine gastric MRI reveals altered Gut-Brain Axis in Functional Dyspepsia: gastric motility is linked with brainstem-cortical fMRI connectivity. Neurogastroenterol. Motil. **34**(10), e14396 (2022). https://onlinelibrary.wiley.com/ doi/10.1111/nmo.14396
13. Van Veen, H., Saul, N., Eargle, D., Mangham, S.: Kepler mapper: a flexible Python implementation of the Mapper algorithm. J. Open Source Softw. **4**(42), 1315 (2019). https://joss.theoj.org/papers/10.21105/joss.01315
14. Hagberg, A., Swart, P.J., Schult, D.A.: Exploring network structure, dynamics, and function using NetworkX, United States, pp. 11–15, January 2008. http:// conference.scipy.org/proceedings/SciPy2008/paper_2/
15. Fortunato, S.: Community detection in graphs. Phys. Rep. **486**(3-5), 75–174 (2010). https://linkinghub.elsevier.com/retrieve/pii/S0370157309002841
16. Desikan, R.S., et al.: An automated labeling system for subdividing the human cerebral cortex on MRI scans into gyral based regions of interest. NeuroImage **31**(3), 968–980 (2006). https://linkinghub.elsevier.com/retrieve/pii/S1053811906000437
17. Zalesky, A., et al.: Whole-brain anatomical networks: does the choice of nodes matter? NeuroImage **50**(3), 970–983 (2010). https://linkinghub.elsevier.com/retrieve/ pii/S1053811909013159
18. Cordasco, G., Gargano, L.: Community detection via semi-synchronous label propagation algorithms. publisher: arXiv Version Number: 1 (2011). https://arxiv.org/ abs/1103.4550
19. Zang, Y., Jiang, T., Lu, Y., He, Y., Tian, L.: Regional homogeneity approach to fMRI data analysis. NeuroImage **22**(1), 394–400 (2004). https://linkinghub. elsevier.com/retrieve/pii/S1053811904000035
20. Rousseeuw, P.J.: Silhouettes: a graphical aid to the interpretation and validation of cluster analysis. J. Comput. Appl. Math. **20**, 53–65 (1987). https://linkinghub. elsevier.com/retrieve/pii/0377042787901257
21. Miri Ashtiani, S.N., et al.: Altered topological properties of brain networks in the early MS patients revealed by cognitive task-related fMRI and graph theory. Biomed. Sig. Process. Control **40**, 385–395 (2018). https://linkinghub.elsevier. com/retrieve/pii/S1746809417302471

P-Count: Persistence-Based Counting of White Matter Hyperintensities in Brain MRI

Xiaoling Hu[1]([✉]), Annabel Sorby-Adams[1], Frederik Barkhof[2,3],
W. Taylor Kimberly[1], Oula Puonti[1,4], and Juan Eugenio Iglesias[1,2,5]

[1] Martinos Center for Biomedical Imaging, MGH and Harvard Medical School,
Boston, USA
`xihu3@mgh.harvard.edu`
[2] Center for Medical Image Computing, University College London, London, UK
[3] Amsterdam University Medical Center, Amsterdam, Netherlands
[4] Danish Research Centre for Magnetic Resonance, Copenhagen University Hospital,
Copenhagen, Denmark
[5] Computer Science and Artificial Intelligence Laboratory, MIT, Cambridge, USA

Abstract. White matter hyperintensities (WMH) are a hallmark of cerebrovascular disease and multiple sclerosis. Automated WMH segmentation methods enable quantitative analysis via estimation of total lesion load, spatial distribution of lesions, and number of lesions (i.e., number of connected components after thresholding), all of which are correlated with patient outcomes. While the two former measures can generally be estimated robustly, the number of lesions is highly sensitive to noise and segmentation mistakes – even when small connected components are eroded or disregarded. In this article, we present *P-Count*, an algebraic WMH counting tool based on persistent homology that accounts for the topological features of WM lesions in a robust manner. Using computational geometry, *P-Count* takes the persistence of connected components into consideration, effectively filtering out the noisy WMH positives, resulting in a more accurate and robust count of true lesions. We validated *P-Count* on the ISBI2015 longitudinal lesion segmentation dataset, where it produces significantly more accurate results than direct thresholding. Our code will be made publicly available upon acceptance.

Keywords: White matter lesions · Persistent homology · Multiple sclerosis

1 Introduction

White matter hyperintensities (WMH) are lesions that appear hyperintense on FLAIR MRI scans. WMH have many possible causes [3], but the two most common are multiple sclerosis (MS) [25] and vascular disorders causing small vessel disease, often leading to stroke [38]. Furthermore, WMH have also been found to be associated with cognitive impairment and Alzheimer's disease (AD) [1]. Therefore, accurate quantification of WMH is highly valuable for the clinical assessment of these diseases and evaluation of potential treatment effect.

Automated lesion segmentation is a key preprocessing step for reproducible, quantitative analysis of WMH – particularly at large scale. Many image segmentation methods have been proposed for WMH. Representative classical methods include: BIANCA [19], which relies on k-nearest neighbor classification; LST-LGA [31], which uses a lesion growth algorithm; LST-LPA [30], which uses supervised logistic regression; atlas-based methods like Lesion-TOADS [32]; dictionary learning algorithms [39]; or unsupervised Bayesian methods that are contrast-adaptive and rely on outlier detection [37], such as BaMoS [33] or SAMSEG-lesion [9].

As in most medical image analysis domains, classical methods have been superseded by deep learning approaches [27]. Many of these methods rely on convolutional neural networks (CNNs) trained in a supervised fashion [6] – possibly equipped with enhancements like positional encoding [18], dedicated patch sampling strategies [20], ensembles [26,28], boundary losses to combat the large class imbalance [23], or longitudinal strategies for jointly exploiting information from multiple timepoints and detect changes [17,24,36]. Attempts have also been made to combine CNNs with ideas from the classical Bayesian literature to achieve resilience against changes in pulse sequence and image resolution [4].

Given segmentations, one can compute several quantitative metrics of interest, which are associated with patient outcomes. One such metric is the total lesion load (also known as "lesion burden"), which corresponds to the total amount of volume segmented as WMH – typically computed from soft segmentations, i.e., by weighting the volume of each voxel by the lesion probability estimated by the automated algorithm at the given location. Lesion load has been shown to correlate with long-term outcomes in MS [29] and stroke [16]. Another important feature of WMH is their spatial distribution. For example, the widespread Fazekas score for grading the amount of WHM in small vessel disease divides lesions into periventricular vs deep white matter [14].

Another metric of interest that can be computed from WMH segmentations is the number of lesions. The appearance of new lesions (or enlargement of existing ones) is used to track the progression of MS in clinical practice, and has been shown to be predictive of disability [7,34,35]. However, counting lesions is an inherently difficult problem, with moderate inter-rater agreement [2,5,41]. While automated segmentation has the potential to curb this variability, counting lesions from a probabilistic segmentation in a robust fashion is not trivial. The standard approach consists of thresholding the lesion probability maps, computing connected components, and filtering out the smallest components. This filtering often relies on morphological erosion, volume thresholding (typically at $\sim 10\,mm^3$ [16]), or geometric criteria (e.g., discarding lesions with less than 3 mm in the major axis [15]). Even when small components are filtered, the lesion count is highly sensitive to the choice of threshold for the soft lesion segmentation, as connected components are created (by splitting) and destroyed as the threshold increases (see Fig. 1). As the experiments below show, this lack of robustness leads to highly variable lesion counts – which is particularly problematic in longitudinal data, as they obscure the real progression of WMH lesions.

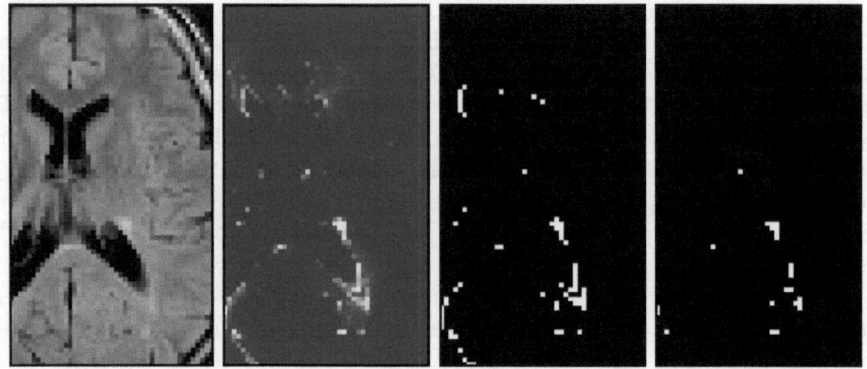

(a) Original slice. (b) Probability map. (c) Seg. under 0.3. (d) Seg. under 0.7.

Fig. 1. Motivation for *P-Count*: lesion counting from an MRI scan (a) based on direct thresholding of the soft probability map (b) is usually noisy and highly sensitive to the choice of the threshold (c, d).

In this article, we present *P-Count*, a novel method for robust lesion count using persistent homology (PH) [12,13]. PH is a topological data analysis tool that has been applied to image segmentation of objects of known topology, both in medical (e.g., vessels, membranes, heart chambers [11,21,22,40]) and natural images [10]. Rather than leveraging PH in a supervised CNN as previous works, here we use it in an unsupervised fashion, to capture the full set of topological changes of the WMH probability map as a function of the threshold. This enables us to count lesions without having to explicitly threshold the probabilities, thus providing more robust estimates.

2   Methods

Our key innovation is to leverage the power of PH to count the number of lesions accurately. Specifically, we capitalize on PH's robustness against noise and its ability to effectively capture the underlying true signal.

2.1   Persistence-Based Counting

We first review the classical watershed algorithm for image segmentation, which is the basis of the proposed method.

Watershed Algorithm. By leveraging topographic information, the watershed algorithm divides an image (2D or 3D) into separated segments. It essentially treats the image as a terrain function (See Fig. 2(a) for a 1D illustration), and identifies basins based on pixel (voxel) intensities. Starting from local minima, the "catchment" basins fill up until region boundaries are reached. Each basin is then labeled as a separate region (c_1 and c_2 in Fig. 2(a)), i.e., as a separate connected component.

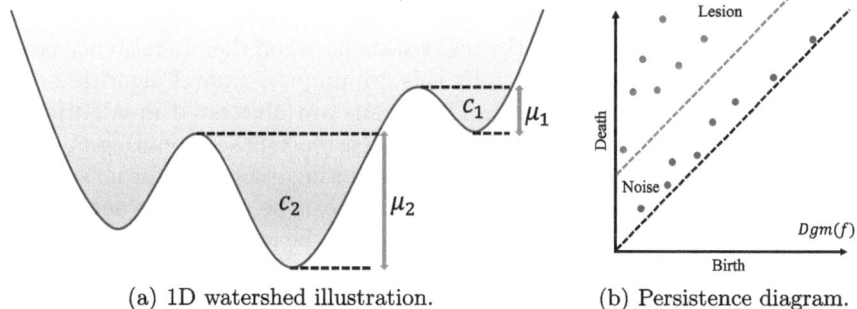

(a) 1D watershed illustration.   (b) Persistence diagram.

Fig. 2. Illustration of the proposed *P-Count*. (**a**) Illustration of the watershed algorithm in 1D. As basins fill up, each is associated with a "lifetime". (**b**) Changes in connected components are captured by the persistence diagram as the threshold increases.

The result of the classical watershed algorithm relies heavily on the local minima. For example, c_1 and c_2 are both counted as valid connected components (i.e., lesions, in our context), regardless of the possibility that they may be noise – which is clearly higher for c_2. This poses a challenge when counting lesions, as every individual local minimum will be included in the result – irrespective of their size or persistence as the threshold varies.

To overcome this issue, we propose to leverage the tool of PH to capture the true lesion signals and suppress the noise. Specifically: instead of counting all the local minima, we seek to distinguish the "real" lesions from noise based on persistence under a progressively increasing threshold. Figure 2(a) shows an example of our intuition in 1D. Every basin (component) is associated with a "lifetime" (μ_1 for c_1, μ_2 for c_2), which is a good indicator of how likely a connected component is to be a real lesion or not. Our method *P-Count* considers the persistence of connected components, effectively filtering out the noisy WMH positives – thus resulting in a more accurate count of true lesions.

Persistence-Based Counting. For each soft probability map, we use a persistence diagram to capture the changes of the topological structure, as illustrated in Fig. 2(b). Each dot in the persistence diagram corresponds to one single connected component, existing at a certain range of threshold values. To distinguish the "real" lesions from noise, we would like to decompose a diagram into "lesion" and "noise" parts.

For a connected component, which corresponds to one dot in the persistence diagram, its lifetime is the persistence of the corresponding dot, defined as the difference between its death and birth time: $per(p) = death(p) - birth(p)$. We note that the persistence of a dot in Fig. 2(b) is equal to the lifetime of its corresponding connected component (e.g., μ_1, μ_2) in Fig. 2(a). Persistence is a good metric indicating how likely a connected component is to be a real lesion or not: the greater the persistence, the longer the connected component exists through the whole "filtration", and the more likely the connected component is to be real. In contrast, the connected components with low persistence are more likely to be "noise".

As a result, we can filter out the real lesions based on their persistence using a predetermined threshold τ. Driven by this, we propose a novel algorithm called *P-Count* for WMH in brain MRI. The details are illustrated in Algorithm 1. As we show in Sect. 3 below, the proposed persistence-based counting algorithm is more robust and results in more accurate counting, especially for noisy inputs. We also show that the choice of threshold: *(i)* can be effectively automatized; *(ii)* has a much lesser impact on the variability of the lesion count than a threshold taken directly on the probability maps; and *(iii)* has very little impact on temporal WMH changes computed from longitudinal scans.

Algorithm 1: *P-Count*

Input: A 3D soft lesion probability map, and a threshold τ
Output: Number of lesions
Definition: $G = (V, E)$ denote a graph; $f(v)$ is the intensity value of node v; $lower\_star(v) = \{(u, v) \in E | f(u) < f(v)\}$; $cc(v)$ is the connected component id of node v.

1: PD $= \emptyset$; Build the proximity graph (6-connectivity) for 3D image;
2: $U = V$ sorted according to $f(v)$; T a sub-graph, which includes all the nodes and edges whose value $< t$.
3: **for** v in U **do**
4:     $t = f(v)$, $T = T + \{v\}$
5:     **for** (u, v) in $lower\_star(v)$ **do**
5:         Assert $u \in T$
6:         **if** $cc(u) = cc(v)$ **then**
6:             Continue
7:         **else**
7:             younger_cc $= \arg\max_{w=cc(u),cc(v)} f(w)$
7:             older_cc $= \arg\min_{w=cc(u),cc(v)} f(w)$
7:             *pers* $= t$-f *(younger_cc)*
8:             **if** pers ¡$= \tau$ **then**
9:                 **for** w in younger_cc **do**
9:                     $cc(w) =$ older_cc
10:                **end for**
11:            **end if**
11:            PD $=$ PD $+ (f($younger_cc$), t)$
12:        **end if**
13:    **end for**
14: **end for**
15: **return** # of lesions $=$ len(cc).

2.2 Optimal Threshold Selection

Let us assume the availability of N training samples, and that the i-th sample has T_i time points. For a set of thresholds $\{\tau_j\}$, sample i has y_{ij1}, y_{ij2}, ..., y_{ijT_i} number of lesions for time point 1, 2, ..., T_i, respectively. The optimal value of threshold depends on the dataset. To find this optimal value, we propose a supervised and an unsupervised approach, depending on whether ground truth labels are available for some scans or not.

Supervised Approach. If ground truth labels are given, we use a supervised approach to select the optimal τ. Let's use $\tilde{y}_{i1}$, $\tilde{y}_{i2}$, ..., $\tilde{y}_{iT_i}$ to denote the ground truth number of lesions for sample i at time point 1, 2, ..., T_i, respectively. Then, we simply pick the threshold that minimizes the sum of absolute errors over all time points of all training samples:

$$\tau^* = arg\,min_\tau \sum_{i=1}^{N} \sum_{t=1}^{T_i} (\tilde{y}_{it} - y_{ijt}(\tau))^2. \tag{1}$$

Unsupervised Approach. If ground truth labels are not available, we utilize an unsupervised method to find the optimal τ. Specifically, we fit a linear model to the number of lesions over time. For sample i under a specific threshold τ_j, we have:

$$\hat{y}_{ijt} = a * t + b + \epsilon_{ijt}, \quad t = 1, 2, ..., T_i. \tag{2}$$

where $\hat{y}_{ijt}$ is the regressed number of lesions at threshold j, time point t for sample i and ϵ_{ijt} models the errors. We use least squares (L2 norm) to fit the linear curve:

$$\tau^* = arg\,min_\tau \sum_{i=1}^{N} \sum_{t=1}^{T_i} (\hat{y}_{ijt} - y_{ijt}(\tau))^2. \tag{3}$$

The optimal τ can be found through Eq. (1) or Eq. (3) under supervised or unsupervised settings, respectively. The optimal τ is then applied to the test set.

3 Experiments and Results

Datasets. To validate the effectiveness of the proposed method, we use the training subset of a longitudinal, multi-modality dataset of WMH (ISBI15 [8]), for which manual longitudinal segmentations are available. The dataset comprises 5 subjects, with 4 or 5 timepoints each. We used the FLAIR scans, resampled to 1 mm isotropic resolution – which is the native resolution of the manual segmentations.

Automated Segmentation. We used SAMSEG-lesion [9] as the automated segmentation method to obtain the soft probabilities from the original scans. We used SAMSEG-lesion because it's adaptive to contrast and therefore generalizes very well to ISBI2015. The output was a soft segmentation of WMH at 1 mm isotropic resolution.

Implementation Details. Our algorithm is computationally expensive, due to the need to monitor connected components at small threshold increases. To reduce its computational burden, we aggressively crop the imaging volumes around the brain, yielding volumes of approximately $80 \times 160 \times 80$ voxels, which allows our single-threaded Python implementation to process a volume in approximately 50 min on a modern desktop.

Baseline. We use direct thresholding (which is the current standard in clinical practice) as the baseline to show the effectiveness of the proposed method. Specifically, we do direct thresholding on the obtained soft probability map, and then count the number of connected components to obtain the number of lesions. We remove connected components with volume under $8 \, \mathrm{mm}^3$ as suggested in [16].

Evaluation. For the baseline method, we gradually increase the threshold from 0.1 to 1 (step size is 0.1), and plot the resulting lesion count vs time curves. For the proposed *P-Count* method, we similarly increase the persistence threshold

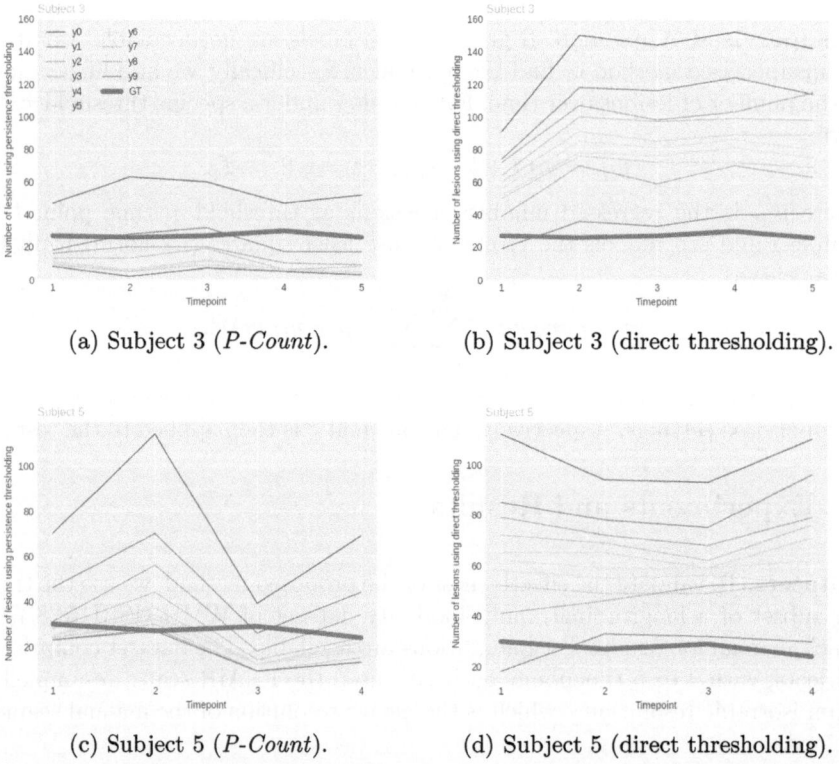

(a) Subject 3 (*P-Count*).   (b) Subject 3 (direct thresholding).

(c) Subject 5 (*P-Count*).   (d) Subject 5 (direct thresholding).

Fig. 3. Lesion count vs time for Subjects 3 and 5 of ISBI2015 using different thresholds (purple = more liberal; red = more conservative), for *P-Count* and direct thresholding. The thick red line corresponds to the ground truth count derived from manual segmentations. (Color figure online)

(τ in Algorithm 1) from 0 to 0.1 (step size is 0.01) to plot the same curves. To evaluate the algorithm quantitatively, we studied the absolute error in lesion count with respect to the ground truth, computed with five-fold cross-validation.

Results. Figure 3 shows the temporal evolution of the number of lesions for two sample subjects in the dataset (Subjects 3 and 5), while Table 1 shows the average errors in lesion counts, for the supervised and unsupervised choices of threshold. Qualitatively speaking, Fig. 3 illustrates two major advantages of *P-Count* with respect to the baseline. First, and most obvious: the error in the lesion count is much lower. This is also apparent from Table 1, with strongly significant reductions in the error rate (despite the small sample size) – particularly for the unsupervised approach. The second improvement is the much lower dependence on the threshold. Direct thresholding of probability maps is very noisy, as further illustrated in Fig. 4(a), which leads to a wide range of trajectories in Fig. 3(a) and Fig. 3(c). This variability is also noticeable in Table 1, where the thresholds obtained by the supervised and unsupervised approaches yield very different error rates. Our method, on the other one, thresholds the *persistence*, yielding curves that are much closer to each other (Figs. 3(b) and 3(d)), segmentations closer to the ground truth (Fig. 4(b)), and errors that are less dependent on the chosen method to determine the threshold (Table 1).

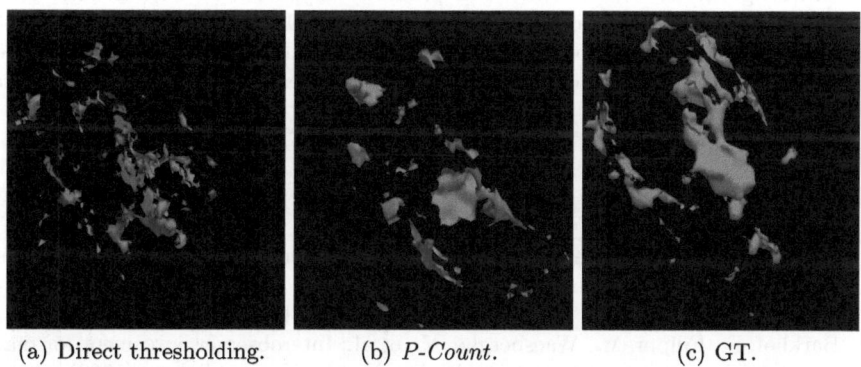

(a) Direct thresholding.   (b) *P-Count*.   (c) GT.

Fig. 4. 3D rendering of WMH for a sample subject. (a) Direct thresholding. (b) *P-Count*. (c) Ground truth. We note that a segmentation like (b) cannot be obtained by thresholding the probability map at any level, as it is based on persistence.

Discussions. This work is a starting point of utilizing persistent homology for lesion counting. In the future, we will expand this research to achieve clinical impact by evaluating various segmentation algorithms beyond SAMSEG-lesion and validating the proposed algorithm across different domains and clinical contexts.

Table 1. Mean absolute errors for lesion count (in number of lesions). The p-values are for two-tailed t-tests comparing the error rates of *P-Count* and direct thresholding.

Method	Mean absolute errors	p-value
Direct thresholding - Unsupervised	23.33	
P-Count - Unsupervised	**17.81**	2.65×10^{-5}
Direct thresholding - Supervised	19.47	
P-Count - Supervised	**10.05**	2.59×10^{-2}

4   Conclusion

We have presented *P-Count*, a PH method for counting WMH that accounts for their topological features in a robust manner. *P-Count* yields much lower errors than the standard thresholding currently used in clinical practice. *P-Count* also has limitations, notably its high computational complexity; future work will seek to address this issue. We believe that *P-Count* has great potential in increasing the accuracy of WMH quantification for the clinical assessment of several diseases and for the evaluation of the effect of treatments.

Acknowledgement. This research was primarily supported by NIH BRAIN grant 1UM1MH130981. Also supported by NIH grants 1RF1MH123195, 1R01AG070988, 1RF1AG080371. OP was supported by a grant from Lundbeckfonden (grant number R360-2021-395). ASA is a recipient of an American Heart Association Postdoctoral Fellowship.

References

1. Alber, J., Alladi, S., Bae, H.J., et al.: White matter hyperintensities in vascular contributions to cognitive impairment and dementia (VCID): knowledge gaps and opportunities. Alzheimer's Dementia: Transl. Res. Clin. Interv. (2019)
2. Barkhof, F., Filippi, M., Waesberghe, V., et al.: Interobserver agreement for diagnostic MRI criteria in suspected multiple sclerosis. Neuroradiology (1999)
3. Barkhof, F., Scheltens, P.: Imaging of white matter lesions. Cerebrovasc. Dis. (2002)
4. Billot, B., Cerri, S., Leemput, V., et al.: Joint segmentation of multiple sclerosis lesions and brain anatomy in MRI scans of any contrast and resolution with CNNs. In: ISBI (2021)
5. Bozsik, B., Tóth, E., Polyák, I., et al.: Reproducibility of lesion count in various subregions on MRI scans in multiple sclerosis. Front. Neurol. (2022)
6. Brosch, T., Tang, L.Y., Yoo, Y., et al.: Deep 3D convolutional encoder networks with shortcuts for multiscale feature integration applied to multiple sclerosis lesion segmentation. TMI (2016)
7. Calabrese, M., Poretto, V., Favaretto, A., et al.: Cortical lesion load associates with progression of disability in multiple sclerosis. Brain (2012)
8. Carass, A., Roy, S., Jog, A., et al.: Longitudinal multiple sclerosis lesion segmentation: resource and challenge. NeuroImage (2017)

9. Cerri, S., Puonti, O., Meier, D.S., et al.: A contrast-adaptive method for simultaneous whole-brain and lesion segmentation in multiple sclerosis. Neuroimage (2021)
10. Chazal, F., Guibas, L.J., Oudot, S.Y., et al.: Persistence-based clustering in Riemannian manifolds. J. ACM (2013)
11. Clough, J.R., Byrne, N., Oksuz, I., et al.: A topological loss function for deep-learning based image segmentation using persistent homology. TPAMI (2020)
12. Edelsbrunner, H., Harer, J., et al.: Persistent homology – a survey. Contemp. Math. (2008)
13. Edelsbrunner, H., Letscher, D., Zomorodian, A.: Topological persistence and simplification. In: FOCS (2000)
14. Fazekas, F., Chawluk, J.B., Alavi, A., et al.: MR signal abnormalities at 1.5 T in Alzheimer's dementia and normal aging. Am. J. Neuroradiol. (1987)
15. Filippi, M., Preziosa, P., Banwell, B.L., et al.: Assessment of lesions on magnetic resonance imaging in multiple sclerosis: practical guidelines. Brain (2019)
16. Georgakis, M.K., Duering, M., Wardlaw, J.M., et al.: WMH and long-term outcomes in ischemic stroke: a systematic review and meta-analysis. Neurology (2019)
17. Gessert, N., Krüger, J., Opfer, R., et al.: Multiple sclerosis lesion activity segmentation with attention-guided two-path CNNs. Comput. Med. Imaging Graph. **84**, 101772 (2020)
18. Ghafoorian, M., Karssemeijer, N., Heskes, T., et al.: Location sensitive deep convolutional neural networks for segmentation of white matter hyperintensities. Sci. Rep. (2017)
19. Griffanti, L., Zamboni, G., Khan, A., et al.: BIANCA (Brain Intensity AbNormality Classification Algorithm): a new tool for automated segmentation of white matter hyperintensities. Neuroimage (2016)
20. Guerrero, R., Qin, C., Oktay, O., et al.: White matter hyperintensity and stroke lesion segmentation and differentiation using convolutional neural networks. NeuroImage Clin. (2018)
21. Hu, X., Li, F., Samaras, D., et al.: Topology-preserving deep image segmentation. In: NeurIPS (2019)
22. Hu, X., Wang, Y., Fuxin, L., et al.: Topology-aware segmentation using discrete Morse theory. In: ICLR (2021)
23. Kervadec, H., Bouchtiba, J., Desrosiers, C., et al.: Boundary loss for highly unbalanced segmentation. In: MIDL (2019)
24. Krüger, J., Opfer, R., Gessert, N., et al.: Fully automated longitudinal segmentation of new or enlarged multiple sclerosis lesions using 3D convolutional neural networks. NeuroImage Clin. (2020)
25. Lassmann, H.: Multiple sclerosis pathology. Cold Spring Harb. Perspect. Med. (2018)
26. Li, H., Jiang, G., Zhang, J., et al.: Fully convolutional network ensembles for white matter hyperintensities segmentation in MR images. NeuroImage (2018)
27. Ma, Y., Zhang, C., Cabezas, M., et al.: Multiple sclerosis lesion analysis in brain magnetic resonance images: techniques and clinical applications. JBHI (2022)
28. Manjón, J.V., Coupé, P., Raniga, P., et al.: MRI white matter lesion segmentation using an ensemble of neural networks and overcomplete patch-based voting. Comput. Med. Imaging Graph. (2018)
29. Popescu, V., Agosta, F., Hulst, H.E., et al.: Brain atrophy and lesion load predict long term disability in multiple sclerosis. J. Neurol. Neurosurg. Psychiatry (2013)
30. Schmidt, P.: Bayesian inference for structured additive regression models for large-scale problems with applications to medical imaging. Ph.D. thesis, LMU (2017)

31. Schmidt, P., Gaser, C., Arsic, M., et al.: An automated tool for detection of flair-hyperintense white-matter lesions in multiple sclerosis. Neuroimage (2012)
32. Shiee, N., Bazin, P.L., Ozturk, A., et al.: A topology-preserving approach to the segmentation of brain images with multiple sclerosis lesions. NeuroImage (2010)
33. Sudre, C.H., Cardoso, M.J., Bouvy, W.H., et al.: Bayesian model selection for pathological neuroimaging data applied to white matter lesion segmentation. TMI (2015)
34. Treaba, C.A., Granberg, T.E., Sormani, M.P., et al.: Longitudinal characterization of cortical lesion development and evolution in multiple sclerosis with 7.0-T MRI. Radiology (2019)
35. Uher, T., Vaneckova, M., Sobisek, L., et al.: Combining clinical and magnetic resonance imaging markers enhances prediction of 12-year disability in multiple sclerosis. Multiple Sclerosis J. (2017)
36. Vaidya, S., Chunduru, A., Muthuganapathy, R., et al.: Longitudinal multiple sclerosis lesion segmentation using 3D convolutional neural networks. In: Proceedings of the 2015 Longitudinal Multiple Sclerosis Lesion Segmentation Challenge (2015)
37. Van Leemput, K., Maes, F., Vandermeulen, D., et al.: Automated segmentation of multiple sclerosis lesions by model outlier detection. TMI (2001)
38. Vermeer, S.E., Hollander, M., van Dijk, E.J., et al.: Silent brain infarcts and white matter lesions increase stroke risk in the general population: the Rotterdam scan study. Stroke (2003)
39. Weiss, N., Rueckert, D., Rao, A.: Multiple sclerosis lesion segmentation using dictionary learning and sparse coding. In: Mori, K., Sakuma, I., Sato, Y., Barillot, C., Navab, N. (eds.) MICCAI 2013. LNCS, vol. 8149, pp. 735–742. Springer, Heidelberg (2013). https://doi.org/10.1007/978-3-642-40811-3_92
40. Wu, P., et al.: Optimal topological cycles and their application in cardiac trabeculae restoration. In: Niethammer, M., et al. (eds.) IPMI 2017. LNCS, vol. 10265, pp. 80–92. Springer, Cham (2017). https://doi.org/10.1007/978-3-319-59050-9_7
41. Zipoli, V., Portaccio, E., Siracusa, G., et al.: Interobserver agreement on Poser's and the new McDonald's diagnostic criteria for multiple sclerosis. Multiple Sclerosis J. (2003)

Outlier Detection in Large Radiological Datasets Using UMAP

Mohammad Tariqul Islam👁 and Jason W. Fleischer$^{(\boxtimes)}$👁

Department of Electrical and Computer Engineering, Princeton University,
Princeton, NJ 08544, USA
{mtislam,jasonf}@princeton.edu

Abstract. The success of machine learning algorithms heavily relies on the quality of samples and the accuracy of their corresponding labels. However, building and maintaining large, high-quality datasets is an enormous task. This is especially true for biomedical data and for metasets that are compiled from smaller ones, as variations in image quality, labeling, reports, and archiving can lead to errors, inconsistencies, and repeated samples. Here, we show that the uniform manifold approximation and projection (UMAP) algorithm can find these anomalies essentially by forming independent clusters that are distinct from the main ("good") data but similar to other points with the same error type. As a representative example, we apply UMAP to discover outliers in the publicly available ChestX-ray14, CheXpert, and MURA datasets. While the results are archival and retrospective and focus on radiological images, the graph-based methods work for any data type and will prove equally beneficial for curation at the time of dataset creation.

Keywords: x-ray · data visualization · data curation · neighbor embedding

1 Introduction

A prominent reason behind the current success of machine learning-based disease detection is the availability of large medical datasets. However, for the machine learning models to be reliable, quality datasets representative of the target population need to be ensured [23]. However, the annotations (and their structured archiving) from automated tools can have errors due to faulty perceptions, interpretations, and human errors [20]. Even if the error rate of the annotator is less than 4%, this can lead to millions of annotation errors per year [3]. This parallels the 3.3% error rate in large computer vision dataset [17]. Thus, there needs to be a better way to identify such errors before they are included in a dataset.

Supplementary Information The online version contains supplementary material available at https://doi.org/10.1007/978-3-031-73967-5_11.

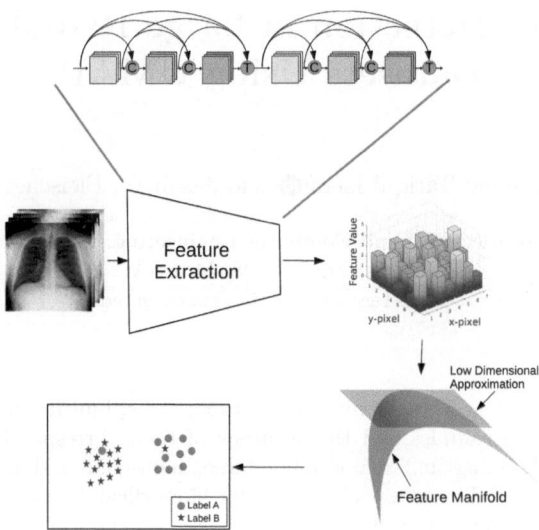

Fig. 1. Schematic of the outlier search algorithm. Image features extracted from a DenseNet-121 neural network are projected onto a low-dimensional space (2-D plane) using UMAP.

For images, the search can be performed visually. However, examining individual images is a daunting task that requires many human hours. A popular alternative is neighbor embedding [8], which can produce a two-dimensional (2-D) cluster plot that can be analyzed visually quickly. (This class of methods is also known as nonlinear dimensionality reduction as the 2-D plot preserves the pairwise similarity, i.e., graph structure, of the original high-dimensional space.) Widely used neighbor embedding algorithms are t-distributed stochastic neighbor embedding (t-SNE) [15] and uniform manifold approximation and projection (UMAP) [16]. UMAP was introduced relatively recently and has become very popular, as this method draws concepts from rich algebraic and topological structures and is computationally fast.

In this paper, we design a UMAP-based visual analytic method for extracting outlier images from large x-ray datasets. We validate our method by analyzing three publicly available and widely used medical image datasets. We show that the method can successfully cluster image features and produce interpretable visualization. We also discover labeling errors and erroneous images that have slipped through the verification process done prior to dissemination. Codes to reproduce the results are available in [13].

2   Related Works

In the literature, the term outlier is often used interchangeably with abnormality and anomaly [5]. Here, we define outliers as images that do not have sufficient signal for final decision-making or do not belong in the dataset due to specification. Generally, outlier detection methods assume an underlying distribution and often model it as normal distribution [7,9], i.e., a data point is an outlier if it is far away from the mean of the fitted distribution. Fritsch et al. [5] used the minimum covariance determinant estimator and its extensions to find outliers (due to motion or registration issues) in neuroimaging data by analyzing principal components. Gang et al. [6] used a t-SNE plot to find outliers from binary lung masks in terms of size variation and segmentation error. Fleischer and Islam [4] employed UMAP on chest x-rays for phenotyping COVID-19 response.

3   System Overview

Following preprocessing (discussed in Supplementary Sect. 1), the major parts of the framework are feature extraction and dimensionality reduction (Fig. 1). To extract features from these images, we employed DenseNet-121 [10] trained on ImageNet [19], a widely used deep neural network architecture designed to efficiently propagate features from earlier layers of a network to deeper layers. Importantly, neural algorithms are usually robust to many variabilities in images by design, and thus can accommodate standard images and outliers on equal terms.

Medical images usually vary in resolution, have different contrast, brightness, and alignment, and often suffer from registration issues. In our framework, the features have been extracted from the final layer (before the softmax layer) of the network, where the features are generally most discriminating. Since we are not using a radiologically pre-trained model, these features generally will not be able to identify individual diseases. Rather, we employ other related labels (e.g., x-ray views, study labels) to examine the datasets. After extracting the features, we employ UMAP [16], to obtain a 2-D approximation of the high-dimensional features.

4   Datasets

We evaluate our approach on three publicly available datasets: ChestX-ray14 [21], CheXpert [11], and Musculoskeletal Radiographs (MURA) [18]. ChestX-ray14 contains 112,120 frontal chest x-ray images from 30,805 unique patients. Images are from posterior-anterior (PA) and anterior-posterior (AP) views. CheXpert dataset contains 224,316 chest x-rays (PA, AP, and Lateral) from 65,240 patients. We used 223,414 JPEG formatted x-rays from the training set of the dataset. MURA dataset contains 40,561 musculoskeletal x-rays from 14,863 studies. Like CheXpert, we used 36,808 x-rays from the training set and utilized study labels, e.g., finger, wrist, hand, forearm, elbow, humerus, and shoulder, for analysis.

5   Experiments

In this section, we start by analyzing embeddings of chest x-rays (ChestX-ray14 and CheXpert datasets) and find different types of outliers. Then, we expand on variations of our method by altering the neural architecture, pre-training dataset, and embedding algorithms. Finally, we discuss outliers in musculoskeletal x-rays of the MURA dataset.

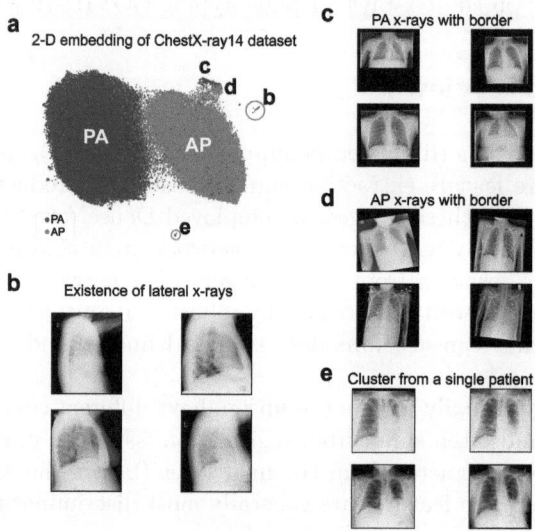

Fig. 2. Outlier detection in the ChestX-ray14 dataset. (a) 2-D embedding. Labeled clusters from (a) are: (b) Lateral x-rays which were not supposed to be in the dataset, (c) PA x-rays with borders, (d) AP x-rays with borders, and (e) cluster from a single patient.

5.1   Analyzing Chest X-Ray Datasets

Lateral X-Rays in ChestX-Ray14: Our 2-D embedding (Fig. 2 (a)) shows two large clusters of PA and AP views, corresponding to the primary topology of the DenseNet features. The embedding also includes a few satellite clusters around these large ones. These satellite clusters around the larger ones occur because the nearest neighbor graph creates a loop (or isolated sub-graph) of common features that are distinct from the rest of the data. In most cases, each of the satellite clusters of x-rays is from a single patient with a unique signature. However, if any specific image features (such as similar artifacts in multiple images) are present in x-rays of different patients, these can create satellite clusters as well. Another interesting structure in Fig. 2 (a) is the protruding region from the AP cluster.

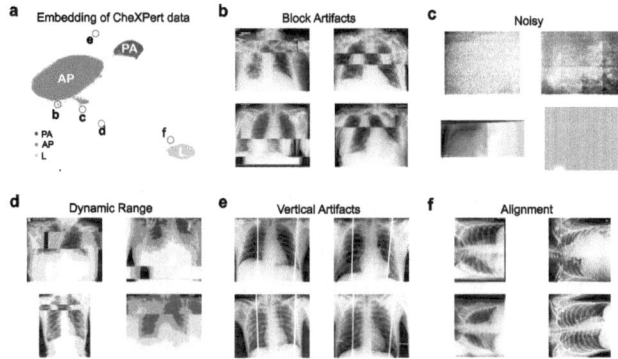

Fig. 3. Outlier detection in the CheXpert dataset. (a) 2-D Embedding. Example images with (b) block artifacts, (c) noise, (d) improper dynamic range, (e) vertical artifacts, and (f) alignment issues.

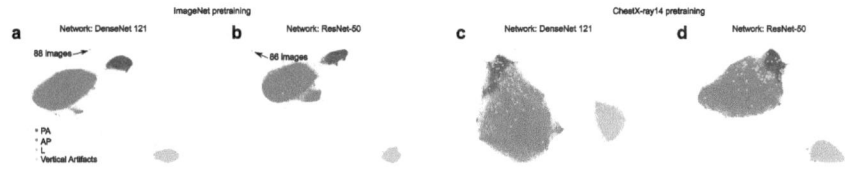

Fig. 4. Embedding of CheXpert dataset using different pre-trained models. DenseNet-121 and ResNet-50 trained on ImageNet (left two) and ChestX-ray14 (right tow) datasets. Each yellow point represents an image with vertical artifact (from cluster e in Fig. 3 (a)) indicating chest x-ray pre-trained models fail to identify these as outliers. (Color figure online)

Representative examples of anomalous clusters are shown in Figs. 2 (b)–(e). The most surprising finding is the existence of some lateral x-rays in the dataset (Fig. 2 (b)), as this dataset is supposed to be composed of frontal chest x-rays only. We found 92 lateral x-rays using our method.

The protruding region from the AP cluster marked c and d (in Fig. 2 (a)), consisting of x-rays with dark borders of PA and AP views, respectively. Finally, cluster (e) shown in Fig. 2 (e) groups 46 x-rays from patient ID 9845, and a single x-ray from patient ID 12562.

Corrupted Images in CheXpert: Figure 3 (a) shows the 2-D embedding of CheXpert dataset. As before, the large PA and AP clusters form the bulk of the mapping. The lateral x-rays also form a separate large cluster. A few of the large satellite clusters (b-f) have been marked by red circles in Fig. 3 (a). Four images from each of the clusters are plotted in Fig. 3 (b)–(e). Figure 3 (b) depicts images with block artifacts, e.g., from poor JPEG compression or accidental splicing. We found 107 such images in this cluster. Figure 3 (c) depicts images that are just noise (19 images). Figure 3 (d) shows images with block artifacts and dynamic

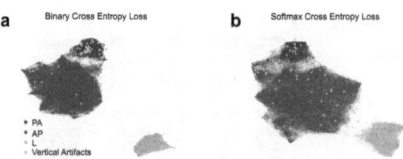

Fig. 5. Pre-training DenseNet-121 models using non-overlapping labels from ChestX-ray14 dataset and embedding of CheXpert dataset using UMAP. Models are trained using (a) binary cross entropy loss and (b) softmax cross entropy loss. Each yellow point represents an image with vertical artifacts (from cluster e in Fig. 3 (a)). (Color figure online)

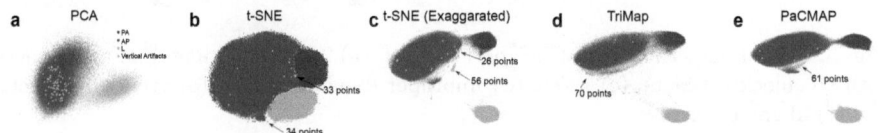

Fig. 6. Embedding of CheXpert dataset using several dimensionality reduction algorithms. (a) PCA, (b) t-SNE, (c) t-SNE (exaggerated), (d) TriMap, and (e) PaCMAP. Each yellow point represents an image with vertical artifacts (from cluster e in Fig. 3(a)). (Color figure online)

range issues (53 images). Figure 3 (e) shows x-rays with vertical artifacts (88 images). Finally, Fig. 3 (f) shows rotated images. This cluster is placed near the large cluster of lateral (L) x-rays. Thus, DenseNet considers rotated x-rays to be more similar to lateral images than upright frontal x-rays.

5.2 Effect of Different Pre-trained Networks

In the previous section, we used DenseNet-121 trained on ImageNet data for feature extraction (Fig. 1). However, using ImageNet may introduce domain shift, since the dataset is not specific to medical images. To assess, we looked at embeddings from different models trained on ImageNet and chest x-rays (Fig. 4).

ImageNet models (Fig. 4 (a, b)) create similar clusters of PA, AP, and L x-rays, whereas Chest-Xray14 models (Fig. 4 (c, d)) merge the PA and AP clusters. The x-ray model embeddings also contain fewer satellite clusters, and their distinctness is lost. Thus, outlier detection fails. Additionally, we highlight the placement of images with vertical artifacts from the cluster (e) of Fig. 3 in the alternate models (yellow points in Fig. 4). While both ImageNet models successfully separate these images (with a few scattered around), all the chest x-ray models fail (where the outliers are inside the large cluster and scattered throughout).

A major difference between models is that ImageNet models use softmax (n-ary) cross entropy loss (for non-overlapping labels), but chest-x-ray models use binary cross entropy loss (for overlapping labels). To resolve this discrepancy, we trained chest x-ray models using non-overlapping labels of the ChestX-ray-14 dataset (for details, see Supplementary Sect. 2). The resulting embeddings (Fig. 5) show a similar characteristic of the chest x-ray models with overlapping labels (Fig. 4 (c, d)), in that the individual views are weakly separated while the specific outlier images fail to form the satellite clusters. For example, the outliers with vertical artifacts scatter within the 2D mapping. The networks consider them as any other x-ray image and ignore the artifacts. This result strengthens the idea that training on ImageNet (or a more general computer vision task) that has broader exposure benefits the discovery of outlier images.

5.3   Comparing Various Embedding Algorithms

To assess the effectiveness of UMAP, we compared it with a few other dimensionality reduction techniques, specifically principal component analysis (PCA), t-SNE, TriMap [1], and PaCMAP [22].

The PCA embedding in Fig. 6 (a) shows the top two directions of largest variances. There are two clusters but the AP and PA views overlap. The nonlinear dimensionality reduction algorithms - t-SNE, UMAP, and variants - discover more features and make the clusters more distinct.

The default t-SNE is tuned to preserve the neighborhood as best as possible. This often causes the individual clusters to spread out and be less compact (Fig. 6 (b)). This behavior is apparent in the PA, AP, and lateral x-ray clusters: the separation among them is minimal, and there is little room for the satellite clusters. However, t-SNE can be tuned to produce a more UMAP-like output. Following the findings of Linderman et al. [14] and Bohm et al. [2], we used an exaggeration factor of 4, which means we applied four times more attractive force than the repulsive force throughout the optimization procedure. The standard early exaggeration factor of 12 was also applied at the start of the optimization. The resulting plot in Fig. 6 (c) largely resolves the PA and AP clusters, but there are fewer satellite clusters than the UMAP output.

Both PaCMAP [22] and TriMap [1] aim to preserve the global structure of the data while keeping the clustering properties of UMAP. PaCMAP modifies the pairwise relation of UMAP by considering different neighborhoods at different scales (near and far), while TriMap achieves this with the triplet constraint. These methods obtain the clustering of PA, AP, and lateral views (Figs. 6 (d, e)), but, similar to exaggerated t-SNE, the satellite clusters are largely absent.

The yellow dots show the placement of x-rays with vertical artifacts from cluster (e) of Fig. 3 for the different embedding algorithms. In all the alternate embedding algorithms, these images are scattered within the other x-ray images. The outlier images fails to be identified and labeled separately, which demonstrates a superior performance of UMAP.

5.4   Extracting Mislabeled X-Rays from MURA

Since the MURA dataset consists of x-rays from different parts of the arm and the shoulder, there is a natural ambiguity in the labels, e.g., both wrist and hand

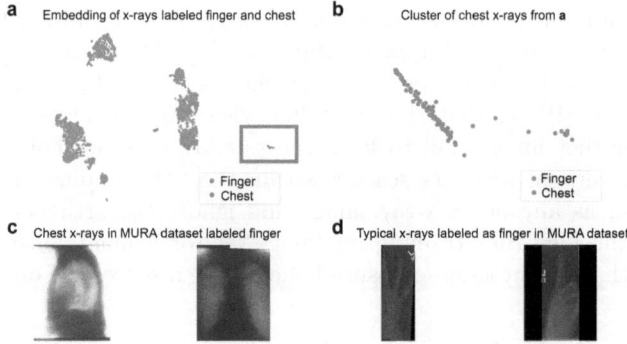

Fig. 7. Embedding of 'finger' x-rays from MURA dataset and 100 chest x-rays from CheXpert dataset using UMAP. (a) Scatter plot of the embedding. The cluster of chest x-rays is marked using a red rectangle. (b) Scatter plot in the red rectangle. (c) two x-rays labeled 'finger' are actually chest x-rays. (d) Typical finger x-rays from the MURA dataset.

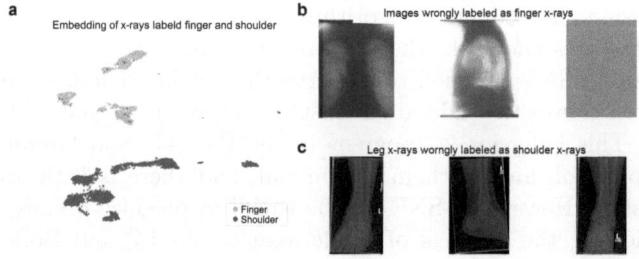

Fig. 8. Embedding of 'finger' and 'shoulder' x-rays from MURA dataset using UMAP. (a) 2-D scatterplot of the embedding. (b) Chest x-ray and non-x-ray images were discovered which are labeled as 'finger' x-rays. (c) Leg x-rays labeled as 'shoulder' x-rays.

x-rays may contain the hand of a person, and shoulder x-rays may contain part of the chest. In such cases, finding mislabeled x-rays by embedding all the images may be sub-optimal. To find outliers more directly, we searched for misclassified images by explicitly using labels of the dataset. The method has two parts: 1) introduce target images with a specific label (preferably from a different dataset than the MURA one); and 2) perform neighbor embedding on the joint dataset.

For example, to look for possible chest x-rays that are falsely classified as finger x-rays, we added 100 chest x-rays from the CheXpert dataset to the 5,106 finger x-rays of MURA. We then applied the UMAP to the composite set (Fig. 7 (a)). As shown in Figs. 7 (b–d), the seeded chest x-rays acted as an attractor for mislabeled images in MURA, with x-rays labeled 'finger' now appearing in the (new) chest cluster. Interestingly, both of these x-rays were from patient 04547 (another 3 from this patient were labeled correctly).

In a different experiment on the MURA dataset, we used 'finger' and 'shoulder' x-rays (Fig. 8). The broad features of the finger and shoulder are easily separable with a few misclassified points. Analyzing 'finger' x-rays misclassified in 'shoulder' clusters, we can find the two chest x-rays labeled as finger (which we found in previous experiment as well) and two images that are just noise/non-x-ray images (Fig. 8 (b)). The latter belongs to patient ID 04687, which uncovers two more outliers (one is an x-ray of keys). The misclassified 'shoulder' labels in the 'finger' cluster reveal three leg x-rays (Fig. 8 (c)).

6   Conclusion

Neighbor embedding algorithms can be an effective tool for summarizing datasets and identifying outlier images. The principle of the method is that the outliers are different from the main data but can have similarities among themselves. Thus, the different outlier types form distinct clusters in the embeddings. Our experiments, using a DenseNet-121 feature extractor and UMAP neighbor embedding method on the ChestX-ray14, CheXpert, and MURA datasets, distinguished different radiological views of chest x-rays, classified differences, and identified wrongly labeled or corrupted images. We further found specific types of outliers by seeding the dataset with target images and performing neighbor embedding.

While this study performed retrospective analysis of large x-ray datasets, outlier curation can be achieved during the initial assembly of the dataset as well. For suspected outliers, the method of seeding data with known labels can be applied. To streamline the process, appropriate reference datasets may be created beforehand. Undoubtedly, cleaner input data will result in cleaner output data. For larger datasets, more accurate results and faster embedding may be achieved by dividing them into smaller subsets and applying better alignment techniques [12]. Finally, since the methods are graph-based and agnostic to the underlying data type, all of the methods here can be applied to arbitrary datasets, including and especially those that are mixed modality.

Acknowledgments. The authors gratefully acknowledge financial support from the Schmidt DataX Fund at Princeton University made possible through a major gift from the Schmidt Futures Foundation.

Disclosure of Interests. The authors have no competing interests to declare that are relevant to the content of this article.

References

1. Amid, E., Warmuth, M.K.: TriMap: large-scale dimensionality reduction using triplets. arXiv preprint arXiv:1910.00204 (2019)
2. Böhm, J.N., Berens, P., Kobak, D.: Attraction-repulsion spectrum in neighbor embeddings. J. Mach. Learn. Res. **23**(1), 4118–4149 (2022)

3. Bruno, M.A., Walker, E.A., Abujudeh, H.H.: Understanding and confronting our mistakes: the epidemiology of error in radiology and strategies for error reduction. Radiographics **35**(6), 1668–1676 (2015)
4. Fleischer, J., Islam, M.T.: Late breaking abstract-identifying and phenotyping COVID-19 patients using machine learning on chest x-rays. Eur. Respir. J. (2020)
5. Fritsch, V., Varoquaux, G., Thyreau, B., Poline, J.B., Thirion, B.: Detecting outliers in high-dimensional neuroimaging datasets with robust covariance estimators. Med. Image Anal. **16**(7), 1359–1370 (2012)
6. Gang, P., et al.: Dimensionality reduction in deep learning for chest X-ray analysis of lung cancer. In: 2018 Tenth International Conference on Advanced Computational Intelligence (ICACI), pp. 878–883. IEEE (2018)
7. Han, S., Hu, X., Huang, H., Jiang, M., Zhao, Y.: ADBench: anomaly detection benchmark. In: Advances in Neural Information Processing Systems, vol. 35, pp. 32142–32159 (2022)
8. Hinton, G., Roweis, S.T.: Stochastic neighbor embedding. In: Advances in Neural Information Processing Systems, vol. 15, pp. 833–840 (2002)
9. Hodge, V., Austin, J.: A survey of outlier detection methodologies. Artif. Intell. Rev. **22**(2), 85–126 (2004)
10. Huang, G., Liu, Z., Van Der Maaten, L., Weinberger, K.Q.: Densely connected convolutional networks. In: Proceedings of the IEEE Conference on Computer Vision and Pattern Recognition, pp. 4700–4708 (2017)
11. Irvin, J., et al.: CheXpert: a large chest radiograph dataset with uncertainty labels and expert comparison. In: Proceedings of the AAAI Conference on Artificial Intelligence, vol. 33, pp. 590–597 (2019)
12. Islam, M.T., Fleischer, J.W.: Manifold-aligned neighbor embedding. In: ICLR 2022 Workshop on Geometrical and Topological Representation Learning (2022)
13. Islam, M.T., Fleischer, J.W.: Codes for outlier detection in large radiological datasets using UMAP (2024). https://github.com/tariqul-islam/Outlier_Detection_UMAP
14. Linderman, G.C., Steinerberger, S.: Clustering with t-SNE, provably. SIAM J. Math. Data Sci. **1**(2), 313–332 (2019)
15. Maaten, L.V.D., Hinton, G.: Visualizing data using t-SNE. J. Mach. Learn. Res. **9**, 2579–2605 (2008)
16. McInnes, L., Healy, J., Melville, J.: UMAP: uniform manifold approximation and projection for dimension reduction. arXiv preprint arXiv:1802.03426 (2018)
17. Northcutt, C.G., Athalye, A., Mueller, J.: Pervasive label errors in test sets destabilize machine learning benchmarks. arXiv preprint arXiv:2103.14749 (2021)
18. Rajpurkar, P., et al.: MURA dataset: towards radiologist-level abnormality detection in musculoskeletal radiographs. In: Medical Imaging with Deep Learning (2018)
19. Russakovsky, O., et al.: ImageNet large scale visual recognition challenge. Int. J. Comput. Vision **115**(3), 211–252 (2015)
20. Waite, S., Scott, J., Gale, B., Fuchs, T., Kolla, S., Reede, D.: Interpretive error in radiology. Am. J. Roentgenol. **208**(4), 739–749 (2017)
21. Wang, X., Peng, Y., Lu, L., Lu, Z., Bagheri, M., Summers, R.M.: ChestX-ray8: hospital-scale chest X-ray database and benchmarks on weakly-supervised classification and localization of common thorax diseases. In: Proceedings of the IEEE Conference on Computer Vision and Pattern Recognition, pp. 2097–2106 (2017)

22. Wang, Y., Huang, H., Rudin, C., Shaposhnik, Y.: Understanding how dimension reduction tools work: an empirical approach to deciphering t-SNE, UMAP, TriMAP, and PaCMAP for data visualization. J. Mach. Learn. Res. **22**(1), 9129–9201 (2021)
23. Yu, K.H., Beam, A.L., Kohane, I.S.: Artificial intelligence in healthcare. Nat. Biomed. Eng. **2**(10), 719–731 (2018)

A Topological Comparison of the Fluorescence Imitating Brightfield Imaging and H&E Imaging

Meiliong Xu[1]([✉]), Nate Anderson[2], Richard M. Levenson[2], Prateek Prasanna[3], and Chao Chen[3]

[1] Department of Computer Science, Stony Brook University, New York, USA
meixu@cs.stonybrook.edu
[2] Department of Pathology and Laboratory Medicine, University of California, Davis, USA
[3] Department of Biomedical Informatics, Stony Brook University, New York, USA

Abstract. Fluorescence Imitating Brightfield Imaging (FIBI) represents an innovative approach in microscopy, providing real-time, non-destructive imaging of tissue without the need for the preparation of thin sections mounted on glass slides. The non-destructive nature of the technology permits tissue preservation for downstream analysis, which makes FIBI a promising alternative to traditional hematoxylin and eosin (H&E) staining in histopathology. Previous research has shown that FIBI can identify morphological features with similar or, in some cases, higher quality compared with H&E images. To comprehensively quantify the advantages and limitations of FIBI in tissue visualization, we propose a novel framework for characterizing the topological difference of FIBI and H&E slide pairs. Experiments are performed on slide pairs of FIBI and H&E imaging of the same tissue area. The proposed approach shows that FIBI can make morphological structures, like vessels, more salient and holds great promise as a complementary technique to H&E, offering novel insights into tissue architecture and potentially improving histopathological diagnostic accuracy.

Keywords: FIBI Imaging · H&E Imaging · Topology

1 Introduction

Histopathological examination of tissue samples is a fundamental aspect of disease diagnosis and research. Hematoxylin and eosin (H&E) staining [12] has long been the gold standard for visualizing tissue morphology and cellular structures [2,3,14–16,18,19,23]. Hematoxylin binds to DNA and RNA, staining nuclei and RNA-rich regions a deep blue or purple hue, providing a clear contrast to the cytoplasm and extracellular matrix. Eosin, an acidic dye, stains mainly cytoplasm and extracellular matrix in varying shades of pink and red. This dual-staining method allows for the detailed visualization and differentiation of

© The Author(s), under exclusive license to Springer Nature Switzerland AG 2025
C. Chen et al. (Eds.): TGI3 2024, LNCS 15239, pp. 122–133, 2025.
https://doi.org/10.1007/978-3-031-73967-5_12

cellular components and tissue structures, supporting the process of arriving at tissue-based diagnoses. However, thin-section H&E has certain limitations, such as limited chromatic range and some difficulties in visualizing extended tubular structures, such as vessels. More generally, there is also the potential for variability in staining quality. These drawbacks have motivated the development of alternative imaging techniques that can address these challenges.

Fluorescence imitating brightfield imaging (FIBI) [11] has emerged as a promising approach, primarily to accelerate the acquisition of histology images from hours to days (with conventional processing) to just a few minutes. However, it can also capture information that can be difficult to assess on conventional H&E-stained slides. FIBI is used to image un-sectioned mm-thick tissue specimens that have been briefly (¡ 1 min) stained with H&E. The method deploys an epifluorescence optical light path to induce a virtual backlight arising from subsurface tissue autofluorescence. Some of this light returns to pass through the stained tissue surface to be collected by a conventional microscope objective and transmitted to a color camera. The images intrinsically resemble brightfield H&E, and do not require significant further processing to be diagnostic. One of the key advantages of FIBI is its ability to provide increased fidelity in visualizing semi-continuous structures such as blood vessels which are typically only irregularly present in thin 5-um sections, potentially enhancing their visibility and the possibility of arriving at more salient representation of such structures. There is a study to investigate FIBI microscopy to evaluate and diagnose feline chronic enteropathy (FCE) [4]. However, to date, there has not been any direct quantitative study of the advantages of FIBI over H&E in terms of structural saliency.

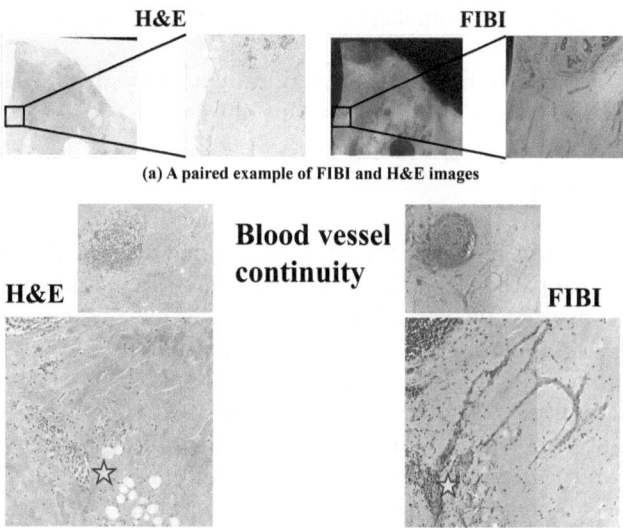

(a) A paired example of FIBI and H&E images

(b) A paired example of areas with small connected vascular structures

Fig. 1. (a) and (b). Paired examples of FIBI and H&E images. The stars in (b) indicate the areas with small connected vascular structures in the tissue samples

To quantitatively compare the structural information captured by FIBI and H&E, we propose the application of tools developed in the field of topological data analysis [7,9], here focusing on persistent homology (PH) [8]. Persistent homology is a powerful tool for quantifying and analyzing the topological features of images across multiple scales. By constructing simplicial complexes and computing topological invariants, persistent homology can capture the essential structural characteristics of an image, including connectedness, holes, and higher-dimensional features. This makes it well-suited for studying the morphological differences between FIBI and H&E-stained images. A paired example of FIBI and H&E images is shown in Fig. 1 (a). A higher magnification comparison is shown in Fig. 1 (b), which documents FIBI's improved ability to detect small connected vascular structures.

In this study, we present a framework based on PH to compare the morphological structural information provided by FIBI and H&E imaging. Our approach involves the preprocessing of the slide pairs, patch cutting, vascular structure segmentation, computation of persistent homology on both types of segmentation results and the analysis of the resulting topological signatures. By quantifying the differences in these signatures, we aim to objectively assess the ability of FIBI to capture and represent tissue morphology compared to traditional H&E staining. Through this investigation, we seek to improve our understanding of the potential of FIBI as a complementary or alternative technique to standard slide imaging in the field of histopathology.

In summary, our main contributions are:

- We design a novel framework that provides a structured approach to compare the topological attributes of FIBI and H&E images.
- We leverage persistent homology to capture and highlight the morphological structures in the two modalities.
- We examine FIBI-H&E slide pairs and evaluate the advantages and limitations of FIBI compared to H&E in terms of topology.

2   Background

Histopathology Imaging. Histopathology imaging technologies are the underpinning of modern pathology by enabling the visualization and analysis of tissue samples at the microscopic level. These technologies have become essential tools for the diagnosis, prognosis, and research of various diseases [13]. The most used histopathology imaging techniques include brightfield microscopy, fluorescence microscopy, complemented by digital pathology whole-slide imaging [21].

Brightfield microscopy, usually based on hematoxylin and eosin (H&E) staining, is the primary imaging modality used in routine clinical practice. H&E staining provides a comprehensive overview of tissue morphology and organization by differentially staining cell nuclei and cytoplasm [12]. Fluorescence microscopy, on the other hand, relies on the use of fluorescent dyes or probes to visualize specific molecules or structures within cells or tissues. Techniques such as immunofluorescence and fluorescence in-situ hybridization (FISH) have enabled the detection and localization of proteins, DNA sequences, and other biomolecules with high specificity and sensitivity [17].

Fluorescence imitating brightfield imaging (FIBI) is a novel histopathology imaging technique that combines some of the properties of brightfield and fluorescence microscopy [10]. FIBI aims to maintain the familiarity and interpretability of brightfield microscopy while complementing it with additional structural and color-based details, the latter arising from additional signals generated by fluorescence excitation [22].

Persistent Homology. In algebraic topology [20], *homology classes* account for topological structures in all dimensions. 0-, 1-, and 2-dimensional structures describe connected components, loops/holes, and cavities/voids, respectively. For binary images, the number of d-dimensional topological structures is called the d-*dimensional Betti number*, β_d.[1] Despite the well-understood topological space for a binary image, the theory does not directly extend to real-world scenarios with continuous, noisy data. For example, in image analysis, we need a principled tool to reason about the topology from a continuous likelihood map. To bridge this gap, the theory of *persistent homology* was invented in the early 2000s [8,25].

Persistent homology has emerged as a powerful tool for analyzing the topology of various kinds of real-world data, including numerical as well as image-based. It quantifies the persistence of topological structures such as connected components, loops, and voids across different scales of filtration. Given an image in the 2D domain $I \subseteq \mathbb{R}^2$, we use a traditional method to generate a likelihood map f. The segmentation map is obtained by thresholding f at a certain threshold c (usually 0.5). We define a *sublevel-set*: $S_c := \{(m, n) \in I \mid f(m, n) \leq c\}$. With all different threshold values sorted in increasing order ($c_1 < c_2 < \cdots < c_n$), we obtain a filtration, i.e., a series of growing sublevel sets: $\varnothing \subseteq S_{c_1} \subseteq S_{c_2} \subseteq \ldots \subseteq S_{c_n} = I$. As the threshold c increases, the topology of the sublevel set changes. New topological structures appear while old ones disappear. Then this information will be aggregated in one tool called persistence diagram. Each point in the persistence diagram corresponds to a topological feature, with its x-coordinate representing the threshold at which the feature appears (birth) and its y-coordinate representing the threshold at which it disappears or merges with another feature (death). The vertical distance of a point from the diagonal line y=x indicates the persistence or lifespan of the corresponding topological feature. Features with high persistence, represented by points far from the diagonal, are often considered more significant and robust to noise, while those close to the diagonal are typically regarded as topological noise. The persistence diagram thus provides a concise summary of the multi-scale topological characteristics of the image.

---

[1] Technically, β_d counts the dimension of the d-dimensional homology group. The number of distinct homology classes/topological structures is exponential to β_d.

3 Methodology

In this section, we introduce our framework in detail. We acquire paired FIBI and H&E slides which enables direct, controlled comparison between these imaging modalities. We make a precise quantification of topological differences attributable to imaging method rather than tissue variability. Topological structures capture fundamental morphological and spatial relationships within tissue architecture, including connectivity of vascular components, spatial organization of tissue structures, and persistent features, providing a robust mathematical framework for objectively comparing FIBI and H&E imaging techniques in representing key tissue characteristics.

We first introduce the acquisition and preprocessing steps of FIBI and H&E slide pairs. Both slides are registered and cut into smaller patches. Smaller image patches facilitate detailed analysis of specific morphological features at higher resolution, reduce computational complexity, and enable granular comparison of local tissue structures. Next, we explain how to segment the morphological structures. Finally, we present how to extract the topological features and aggregate them to facilitate the analysis. The overview pipeline is shown in Fig. 2.

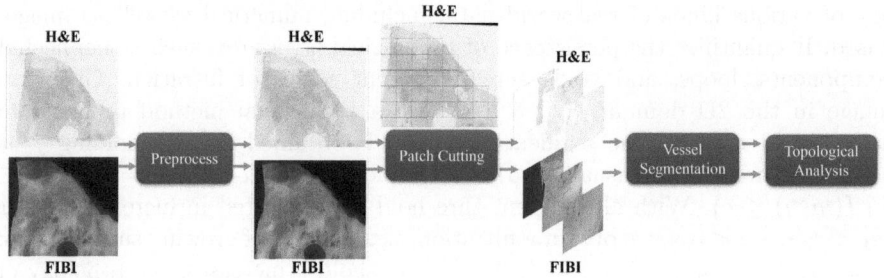

Fig. 2. Overview pipeline of the framework.

Data Acquisition. The FIBI and H&E image pairs can be created in two ways, both of which allow for close correlation between the two modes. In the first, thick tissue specimens are imaged using FIBI, and then that piece of tissue is paraffin-embedded and thin-sectioned to create a slide, with attention paid to keep the distance between the surface imaged with FIBI and that ending up on the glass slide as small as possible (a few 10's of microns). Alternatively, the FIBI image can be acquired from the face of the paraffin block after a slide has been prepared, again helping to minimize the difference between tissue structures imaged in both modes.

Preprocessing. Preprocessing procedures are implemented to ensure equitable comparison between FIBI and H&E-stained slide pairs. These steps include image resizing for spatial normalization, rotation correction for tissue alignment, and affine transformation-based image registration. The latter utilizes manually selected three control points to optimize transformation parameters, thereby achieving precise overlay and spatial alignment between modalities. While these preprocessing steps establish a robust foundation for comparative analysis, they are anticipated to become unnecessary as the methodology matures. The resulting preprocessed slide pairs, comprising one FIBI slide (S_F) and one H&E slide (S_H) of identical tissue areas, undergo subsequent glass detection and exclusion. This process employs color-based thresholding to identify and remove non-diagnostically relevant regions. Following these preparatory steps, the sliding window method [18] is applied. This technique segments the H&E slide into n patches, denoted as $S_H = \{h_1, h_2, ..., h_n\}$. Corresponding patches from S_F are then extracted based on the coordinates derived from S_H.

Vessel Segmentation. We focus on one kind of representative morphological structure: vascular structures. As we do not have the ground truth for FIBI and H&E, we adopt a traditional method to segment vessels: multi-scale Hessian filter [24].

The multi-scale Hessian-based measure is a mathematical approach that leverages the second-order derivatives of the image intensity to identify tubular structures, such as blood vessels. The Hessian matrix, which contains the second-order partial derivatives of the image intensity, provides valuable information about the local curvature and orientation of structures within the image.

To segment blood vessels using the multi-scale Hessian-Based Measure, we first calculate the Hessian matrix H of an image I. The Hessian matrix of an image at a location $p = (x, y)$ is defined as the matrix of second-order partial derivatives:

$$H(I, \sigma)(p) = \begin{bmatrix} \frac{\partial^2 I}{\partial x^2} & \frac{\partial^2 I}{\partial x \partial y} \\ \frac{\partial^2 I}{\partial y \partial x} & \frac{\partial^2 I}{\partial y^2} \end{bmatrix} \tag{1}$$

where σ is the scale parameter of the Gaussian kernel used for smoothing the image and reducing the noise. The Gaussian smoothed image I_σ is given by: $I_\sigma(p) = I(p) * G(p, \sigma)$, with $G(p, \sigma)$ being the Gaussian kernel defined as $G(p, \sigma) = \frac{1}{2\pi\sigma^2} e^{-\frac{x^2+y^2}{2\sigma^2}}$. The eigenvalues λ_1 and λ_2 (with $|\lambda_1| \leq |\lambda_2|$) of the Hessian matrix provide crucial information about the local structure of the image. The vesselness measure V at a given scale σ is formulated to enhance tubular structures and suppress other structures given by:

$$V(\sigma) = \begin{cases} 0 & \text{if } \lambda_2 > 0 \\ \exp\left(-\frac{R_B^2}{2\beta^2}\right)\left(1 - \exp\left(-\frac{S^2}{2c^2}\right)\right) & \text{otherwise} \end{cases} \tag{2}$$

where $R_B = \frac{|\lambda_1|}{\lambda_2}$, $S = \sqrt{\lambda_1^2 + \lambda_2^2}$. β and c are parameters controlling the sensitivity of the measure to blob-like structures and the scale of the vessels, respectively. To capture vessels of varying sizes, the vesselness measure is computed at multiple scales σ_i: $V_{\text{multi-scale}} = \max_i V(\sigma_i)$, where σ_i spans a range of scales.

Finally, the binary segmentation map B of image I is obtained by threshold-ing the multi-scale vesselness measure. For both FIBI and H&E patches, the multi-scale Hessian filter is applied to obtain vessel segmentation maps:

$$B_F = \{b_f^1, b_f^2, ..., b_f^n\} \ B_H = \{b_h^1, b_h^2, ..., b_h^n\} \tag{3}$$

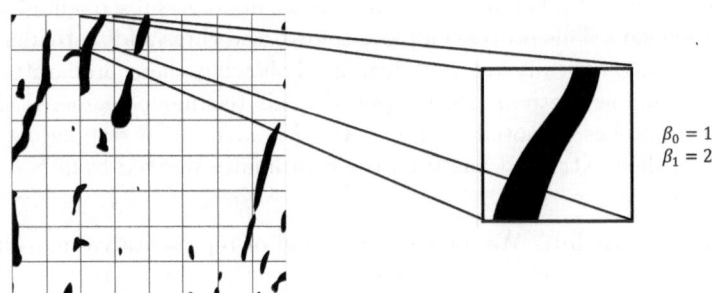

Fig. 3. Interpretation of Betti numbers in a vessel segmentation patch: $\beta_0 = 1$ (number of connected components), $\beta_1 = 2$ (number of loops or circular structures).

Topological Analysis. In this section, we will introduce how to capture the topological features. Initially, the segmentation map is divided into smaller, non-overlapping patches to facilitate localized analysis. Subsequently, a block is added to each patch to enable the calculation of one-dimensional topological structures, as shown in Fig. 3. This step is essential for capturing the connectivity and branching patterns of the vasculature. Following this, we apply persistent homology, a robust tool from topological data analysis, to each smaller patch. Ini-tially, we compute the distance transform of each patch [5], a crucial step in cap-turing the topological information of the segmentation map. The distance trans-form provides a continuous representation of the binary segmentation, encoding not only the presence of structures but also their spatial relationships and rela-tive distances. This transformation is essential as it allows for a more nuanced analysis of the topological features, capturing gradual changes and preserving geometric information that would be lost in a purely binary representation.

Subsequently, we calculate the persistent homology based on these distance transforms. The output of this computation is a persistence diagram for each FIBI and H&E patch pair, which succinctly represents the birth and death of topological features as the threshold varies:

$$Dgm(b_f^i) = PH(DT(b_f^i)) \ Dgm(b_h^i) = PH(DT(b_h^i)) \tag{4}$$

Then, to better demonstrate the topological difference between FIBI and H&E, we calculate the Wasserstein Distance [6] between two diagrams. Given two diagrams $Dgm(g)$ and $Dgm(h)$, the p-th Wasserstein distance between them is defined as:[2]

$$W_p(Dgm(g), Dgm(h)) = \left(\underset{\gamma \in \Gamma}{inf} \sum_{x \in Dgm(g)} ||x - \gamma(x)||^p \right)^{\frac{1}{p}}$$

where Γ represents all bi-jections from $Dgm(g)$ to $Dgm(h)$.

Besides Wasserstein distances, we also derive the Betti numbers for each smaller patch. As illustrated in Fig. 3, the β_0 and β_1 of a representative smaller patch are 1 and 2, respectively, indicating the number of connected components and holes in the local structure. By systematically computing these Betti numbers for each patch, we obtain a comprehensive characterization of the local topological properties of the vessel segmentation map, effectively quantifying the complexity and connectivity of the segmented structures at various scales.

One limitation of the proposed approach is the fixed patch size. One appealing property of TDA is its ability to capture topological structures of multiple scales. By fixing the patch size (512×512 pixel^2), we essentially only capture structures up to this scale. We will leave a multi-scale solution to future work.

4   Experiments

4.1   Dataset and Settings

Eight whole slide image pairs of different tissue types are used to evaluate the topological differences between FIBI and H&E imaging, each consisting of FIBI and H&E images of the same tissue area. Initially, we use a patch size of 4096×4096 for vessel detection and segmentation. Afterward, when carrying out topological analysis, we use a smaller patch size (512×512).

Quantitative Results. Based on the topological analysis on 12416 patches presented in Fig. 4, we can infer that FIBI demonstrates enhanced capabilities in modeling morphological structures, particularly vessels, compared to traditional Hematoxylin and Eosin (H&E) staining. The histograms of 0-dimensional and 1-dimensional Betti numbers provide quantitative evidence for this assertion. In the 0-dim Betti number analysis, which represents connected components, FIBI exhibits a broader distribution with a higher median value compared to H&E, suggesting a more nuanced capture of distinct structural elements. Similarly, the 1-dim Betti numbers, corresponding to loops or holes in the topological structure, exhibit a more diverse distribution for FIBI, suggesting a superior ability to detect and represent complex vascular networks.

Furthermore, the histograms of Wasserstein distances (Fig. 5) provide a quantitative measure of the topological differences between FIBI and H&E image

---

[2] For ease of exposition, we change the original formulation and use the 2-norm instead of infinity norm for $||x - \gamma(x)||$. The difference is bounded by a $\sqrt{2}/2$ constant factor.

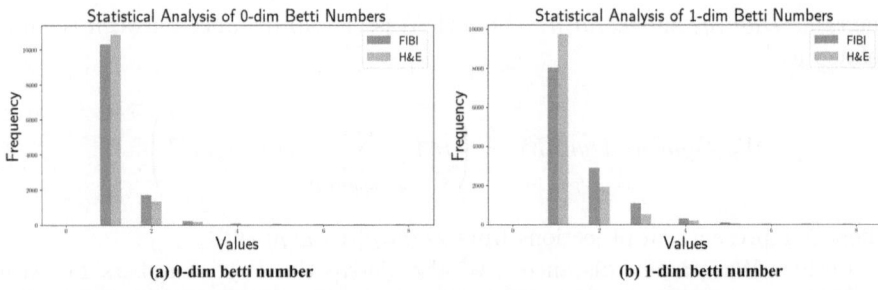

(a) 0-dim betti number   **(b) 1-dim betti number**

Fig. 4. The statistical analysis of topological properties.

pairs. The 0-dim topological difference histogram shows a spread distribution, indicating significant variations in the capture of connected components between the two imaging modalities. The 1-dim topological difference histogram is particularly telling, with a high frequency of non-zero Wasserstein distances, emphasizing FIBI's enhanced capability in representing higher-order topological features such as loops in vascular structures.

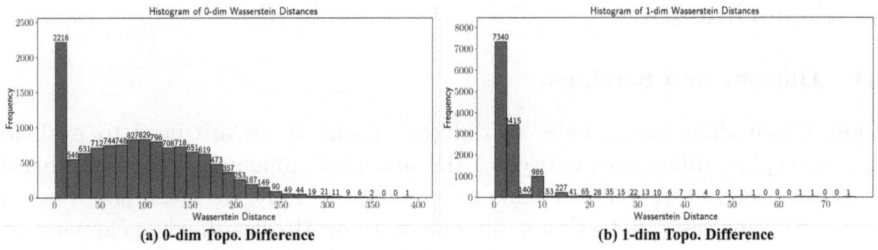

(a) 0-dim Topo. Difference   **(b) 1-dim Topo. Difference**

Fig. 5. The topological difference of FIBI and H&E image pairs.

The quantitative analysis of Betti numbers shown in Table 1 reveals significant topological differences between FIBI and H&E imaging techniques. For both 0-dimensional and 1-dimensional Betti numbers, FIBI consistently demonstrates higher values compared to H&E, demonstrating that FIBI can provide improved visualization and quantification of morphological features.

Table 1. Quantitative Results of 0 and 1-dim Betti numbers.

	FIBI	H&E	p-value
0-dim Betti Number	1.2845 ± 0.8271	1.1742 ± 1.1509	4.6279^{-18}
1-dim Betti Number	1.5215 ± 0.6754	1.3027 ± 0.8560	1.0735^{-9}

This enhanced sensitivity to topological variations suggests that FIBI can provide a more detailed and potentially more accurate representation of vessel morphology and other critical tissue structures. These findings underscore FIBI's potential as a powerful complementary technique to H&E in histopathological analysis, offering improved visualization and quantification of morphological features that may be less apparent in traditional staining methods.

Qualitative Results. The rich vasculature structures visible in FIBI provides new opportunity for vessel-informed tumor microenvironment analysis. We apply cell detection models MCSpatNet [1] to detect cells from these images. Three types of cells are detected, lymphocytes in blue, tumor/epithelial cells in red, and stromal cells in green. As we can see in Fig. 6, we can segment the vessel structures very clearly in the FIBI patch as shown in Fig. 6(a). This is due to the higher contrast of morphological structures, as well as the unique 2.5D imaging in FIBI. Meanwhile, we can detect cells from both FIBI and H&E imaging Fig. 6(a)(b). Since in this case, the model is pre-trained on H&E, we can use the cell detection results from H&E overlaid with vessel detection result from FIBI (Fig. 6(c)) for further analysis. This demonstrates that FIBI can be a promising tool, offering novel insights into tissue architecture and potentially improving histopathological diagnostic accuracy.

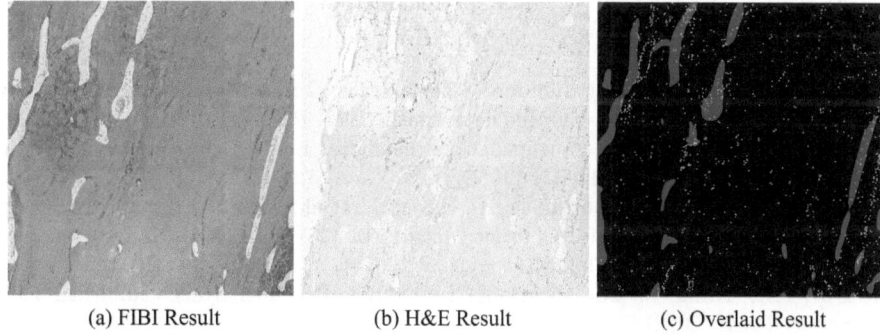

(a) FIBI Result   (b) H&E Result   (c) Overlaid Result

Fig. 6. The results of cell detection and vessel segmentation on FIBI and H&E patch pairs. (a) the result on FIBI; (b) the result on H&E; (c) the overlaid result. (Color figure online)

5   Conclusion

In conclusion, this study presents a novel framework for quantitatively assessing the topological differences between FIBI and traditional hematoxylin and eosin (H&E) preparations in histopathology. Using topological analysis, we capture the structural insights provided by FIBI due to the visibility of vasculature structures. This is a good case study to show the power of TDA in real-world applications with its unique structural quantification power. The result demonstrates the strong promise of FIBI as a novel imaging technique. While FIBI may not entirely replace H&E in all diagnostic scenarios, its capacity to offer novel insights into tissue architecture suggests significant potential for improving histopathological diagnostic accuracy. Future research should focus on expanding the application of this framework to a broader range of tissue types and pathological conditions, and in particular, explore the use of FIBI images to compare topologies associated with disease severity and outcome.

Acknowledgement. We thank the anonymous reviewers for their constructive feedback. The research was partially supported by NIH R21CA258493-02S1 and R33CA278544-01S1.

References

1. Abousamra, S., et al.: Multi-class cell detection using spatial context representation. In: Proceedings of the IEEE/CVF International Conference on Computer Vision (2021)
2. Abousamra, S., Gupta, R., Kurc, T., Samaras, D., Saltz, J., Chen, C.: Topology-guided multi-class cell context generation for digital pathology. In: Proceedings of the IEEE/CVF Conference on Computer Vision and Pattern Recognition (2023)
3. Alsaleh, L., et al.: Spatial transcriptomic analysis reveals associations between genes and cellular topology in breast and prostate cancers. Cancers **14**(19), 4856 (2022)
4. Au Yeung, S., et al.: Utility of fluorescence imitating brightfield imaging microscopy for the diagnosis of feline chronic enteropathy. Vet. Pathol. **60**(1), 52–59 (2023)
5. Borgefors, G.: Distance transformations in digital images. Comput. Vis. Graph. Image Process. **34**(3), 344–371 (1986)
6. Cohen-Steiner, D., Edelsbrunner, H., Harer, J., et al.: Lipschitz functions have Lp -stable persistence. Found. Comput. Math. **10**, 127–139 (2010). https://doi.org/10.1007/s10208-010-9060-62010
7. Dey, T.K., Wang, Y.: Computational Topology for Data Analysis. Cambridge University Press (2022)
8. Edelsbrunner, H., Letscher, D., Zomorodian, A.: Topological persistence and simplification. Discrete Comput. Geom. **28**, 511–533 (2002). https://doi.org/10.1007/s00454-002-2885-2
9. Edelsbrunner, H., Harer, J.L.: Computational Topology: An Introduction. American Mathematical Society (2010)
10. Fereidouni, F., Levenson, R.M.: Fluorescence imitating brightfield imaging, US Patent, 11808703B2 (2023)
11. Fereidouni, F., Morningstar, T., Borowsky, A., Levenson, R.: FIBI: a direct-to-digital microscopy approach for slide-free histology. In: Medical Imaging 2022: Digital and Computational Pathology, vol. 12039, p. 1203903. SPIE (2022)
12. Fischer, A.H., Jacobson, K.A., Rose, J., Zeller, R.: Hematoxylin and Eosin Staining of Tissue and Cell Sections. Cold Spring Harbor Protocols (2008)
13. Gu, J., Taylor, C.R.: Practicing pathology in the era of big data and personalized medicine. Appl. Immunohistochem. Mol. Morphol. (2014)

14. Hasan, M., Xiaoling, H., Abousamra, S., Prasanna, P., Saltz, J., Chen, C.: Semi-supervised contrastive VAE for disentanglement of digital pathology images. In: MICCAI. Springer (2024)
15. Huang, W., Xiaoling, H., Abousamra, S., Prasanna, P., Chen, C.: Hard negative sample mining for whole slide image classification. In: MICCAI. Springer (2024)
16. Kapse, S., et al.: SI-MIL: taming deep MIL for self-interpretability in gigapixel histopathology. In: Proceedings of the IEEE/CVF Conference on Computer Vision and Pattern Recognition (2024)
17. Levsky, J.M., Singer, R.H.: Fluorescence in situ hybridization: past, present and future. J. Cell Sci. **116**(14), 2833–2838 (2003)
18. Li, B., Li, Y., Eliceiri, K.W.: Dual-stream multiple instance learning network for whole slide image classification with self-supervised contrastive learning. In: CVPR, pp. 14318–14328 (2021)
19. Li, C., Xiaoling, H., Abousamra, S., Meilong, X., Chen, C.: Spatial diffusion for cell layout generation. In: MICCAI. Springer (2024)
20. Munkres, J.R.: Elements of Algebraic Topology. CRC Press (1984)
21. Niazi, M.K.K., Parwani, A.V., Gurcan, M.N.: Digital pathology and artificial intelligence. Lancet Oncol. **20**(5), e253–e261 (2019)
22. Volpi, C.C., Gualeni, A.V., Pietrantonio, F., Vaccher, E., Carbone, A., Gloghini, A.: Bright-field in situ hybridization detects gene alterations and viral infections useful for personalized management of cancer patients. Expert Rev. Mol. Diagn. **18**(3), 259–277 (2018)
23. Xu, M., Hu, X., Gupta, S., Abousamra, S., Chen, C.: TopoSemiSeg: enforcing topological consistency for semi-supervised segmentation of histopathology images. arXiv preprint arXiv:2311.16447 (2023)
24. Yang, J., et al.: Improved Hessian multiscale enhancement filter. Biomed. Mater. Eng. **24**(6), 3267–3275 (2014)
25. Zomorodian, A., Carlsson, G.: Computing persistent homology. In: Proceedings of the Twentieth Annual Symposium on Computational Geometry, pp. 347–356 (2004)

Topological Analysis of Seizure-Induced Changes in Brain Hierarchy Through Effective Connectivity

Anass B. El-Yaagoubi[1]($\boxtimes$) ⓘ, Moo K. Chung[2] ⓘ, and Hernando Ombao[1] ⓘ

[1] King Abdullah University of Science and Technology (KAUST),
Statistics Program, Thuwal 23955, Saudi Arabia
{anass.bourakna,hernando.ombao}@kaust.edu.sa
[2] Department of Statistics, University of Wisconsin-Madison, Medical Science Center
4725, 1300 University Ave, Madison, WI 53706, USA
mkchung@wisc.edu

Abstract. Traditional Topological Data Analysis (TDA) methods, such as Persistent Homology (PH), rely on distance measures (e.g., cross-correlation, partial correlation, coherence, and partial coherence) that are symmetric by definition. While useful for studying topological patterns in functional brain connectivity, the main limitation of these methods is their inability to capture the directional dynamics - which are crucial for understanding effective brain connectivity. We propose the Causality-Based Topological Ranking (CBTR) method, which integrates Causal Inference (CI) to assess effective brain connectivity with Hodge Decomposition (HD) to rank brain regions based on their mutual influence. Our simulations confirm that the CBTR method accurately and consistently identifies hierarchical structures in multivariate time series data. Moreover, this method effectively identifies brain regions showing the most significant interaction changes with other regions during seizures using electroencephalogram (EEG) data. These results provide novel insights into the brain's hierarchical organization and illuminate the impact of seizures on its dynamics.

Keywords: Hodge Decomposition · Topological Data Analysis · Time Series Analysis · Effective Brain Connectivity · Seizure EEG Data

1 Introduction

Over the past two decades, Topological Data Analysis (TDA) has become a valuable tool in various fields, providing new insights into complex data structures [1,5,6,8,9]. In biology, TDA has helped clarify genetic and evolutionary processes [19], while in materials science, it has improved predictions of properties in crystalline compounds [16]. In finance, TDA has offered a novel perspective on the dynamics of financial markets during crises [14]. In neuroscience, the use of Persistent Homology (PH), a key technique within TDA, has revealed

C. Chen et al. (Eds.): TGI3 2024, LNCS 15239, pp. 134–145, 2025.
https://doi.org/10.1007/978-3-031-73967-5_13

new aspects of the topological organization of brain dependence networks. By analyzing multivariate brain signals, researchers have uncovered complex patterns and structures that deepen our understanding of brain connectivity and functionality [2,10,17].

However, traditional uses of PH often involve Vietoris-Rips filtrations from *undirected* brain dependence networks. A significant limitation of this approach is its inability to capture the directionality of connectivity between nodes in a brain network [11]. It is well-established that node dependencies within a brain network exhibit asymmetry, where the influence exerted by node A on node B can differ significantly from the influence exerted by node B on node A. This asymmetry is crucial for understanding effective brain connectivity, which involves directional information flow, from one region into another, that often varies markedly in its reverse path, thus delineating specific neural pathways [12,13,20,23].

In the study of epilepsy, detecting changes in brain connectivity is a crucial task. Research indicates that epileptic seizures is associated with significant alterations in interactions between brain regions in a network. These changes vary depending on whether the seizure is generalized or originates from a specific focal region and spreads across the brain's hemispheres [4,7]. Traditional TDA methods, which primarily assess functional connectivity, often fall short in accurately characterizing these dynamic changes in effective brain connectivity.

Causal inference is the process of learning a causal model from observational data, which is essential in cognitive neuroscience for understanding how brain regions interact and how these interactions may change due to disorders [25]. Experimental interventions, such as optogenetics, are the gold standard in causal discovery [18,25], allowing researchers to manipulate neuron activity and observe changes in neural interactions. While these methods provide definitive evidence of causal relationships, they are impractical and ethically challenging in human studies. Therefore, our study will focus on identifying and quantifying causal associations from observational data alone.

To better understand dynamic changes in brain connectivity during epileptic seizures, we developed the Causality-Based Topological Ranking (CBTR) approach. This method combines Causal Inference (CI) and Hodge Decomposition (HD) to identify causal pathways and hierarchically rank brain regions based on their mutual influence over time. CBTR assesses dependencies while accounting for parent nodes, thereby minimizing confounding effects. Our analysis of EEG data from seizure-affected subjects identified the brain regions most significantly impacted for each individual.

2   Methods

The goal is to develop a method for ranking brain regions (derived from the network of electroencehphalogram data) according to effective connectivity using a two-step process. First, causal inference techniques are employed to map the hierarchical structure of brain connectivity, with a focus on the effects of seizures.

Next, Hodge Decomposition is applied to net flow of information between brain regions, quantifying their mutual influence.

2.1   Causal Inference of the Brain's Hierarchical Structure

Understanding the dynamics of the hierarchical structure of the brain's effective connectivity is crucial for comprehending the impact of seizures. In an EEG network, a channel is considered to be at the top of the hierarchy if it tends to exert greater influence on other nodes, whereas a brain channel is at the bottom if it tends to be influenced more by other nodes. This hierarchical structure is determined by analyzing the causal relationships and time-delayed dependencies among the channels. This is illustrated in Fig. 1. Assessing changes in brain hierarchy allows us to quantify the effects of seizures on each specific EEG channel, providing deeper insights into how seizures alter brain connectivity.

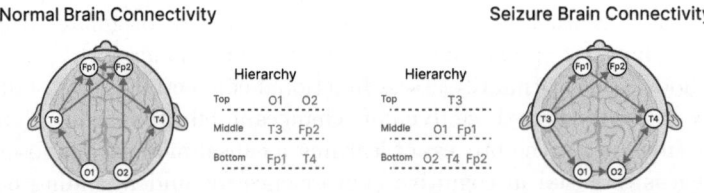

Fig. 1. Potential hierarchy of the brain in normal (Left) and seizure (Right) conditions. In the normal state, the hierarchy is O1 (left occipital), O2 (right occipital) followed by T3 (left temporal), Fp2 (right frontal pole), and then T4 (right temporal), Fp1 (left frontal pole). During a seizure, the order changes to T3, O1, Fp1 followed by T4, O2, Fp2.

To estimate effective brain connectivity, we used the PCMCI (Peter and Clark Momentary Conditional Independence) method, a robust causal inference technique suited for high-dimensional time series data [21,22]. PCMCI involves a two-stage approach: first, the PC algorithm identifies potential causal parents for each variable, reducing dimensionality and creating a preliminary causal graph. Second, the Momentary Conditional Independence (MCI) test assesses the significance of these dependencies, considering both contemporaneous and time-lagged relationships while controlling for false positives.

The outputs of the PCMCI method include a causal graph that highlights the most significant causal links between brain regions at various lags, with each link annotated by its corresponding statistical significance and strength. This causal graph forms the foundation for our subsequent analyses. We estimate effective connectivity as a weighted directed network, where the weight of each link ($\mathbf{W}_{p,q}$) between regions p and q is derived from the significance level (p-value) associated with the causal interaction from q to p, as defined in Eqs. 1, 2.

$$\mathbf{W}_{p,q} = 1 - \overline{p}_{p,q}, \tag{1}$$

$$\overline{p}_{p,q} = \frac{1}{K} \sum_{k=0}^{K} p_{p,q}(k), \tag{2}$$

where $p_{p,q}(k)$ is the p-value corresponding to the MCI test from variable q to p at lag k. The estimated effective brain connectivity network is decomposed into symmetric ($\mathbf{W}_s$) and anti-symmetric ($\mathbf{W}_a$) components.

$$\mathbf{W} = \mathbf{W}_s + \mathbf{W}_a \tag{3}$$

where $\mathbf{W}_s = \frac{1}{2}(\mathbf{W} + \mathbf{W}')$ represents mutual causal influence and $\mathbf{W}_a = \frac{1}{2}(\mathbf{W} - \mathbf{W}')$ represents net causal influence. Since brain dependencies can be asymmetric, this decomposition is relevant as it separates the connectivity network into meaningful symmetric and anti-symmetric components. The symmetric component, $\mathbf{W}_s$, captures bidirectional interactions, reflecting mutual influences critical for cooperative neural processes. The anti-symmetric component, $\mathbf{W}_a$, highlights the predominant direction of causal influence, revealing the hierarchical structure and directional information flow within the brain network. This approach provides a nuanced understanding of neural activities and identifies key regions driving effective connectivity. This decomposition is attractive to neuroscientists - it is an essential tool to deepen our understanding of the hierarchical structure of the brain. The next step involves building the ranking of brain regions using Hodge Decomposition, which will allow us to quantify and rank the hierarchical influence of each region within the network.

2.2   Proposed Method for Ranking Brain Regions

Once effective connectivity is estimated and the net influence $\mathbf{W}_a$ is derived, our proposed method will rank brain regions by assigning a score to each node or brain channel based on their influence within the network. Our method is based on the HodgeRank approach, as described in [15], which has previously been utilized to rank entities such as movies and football teams. A higher ranking score indicates a region that influences other regions more than it is influenced, while a lower score suggests a region that is more influenced by others. This concept is illustrated in Fig. 2.

To achieve the most meaningful ranking (i.e., one that minimizes violations of mutual net causal influence), we apply Hodge Decomposition to the anti-symmetric component of the connectivity network. Since $(\mathbf{W}_a)_{p,q} = -\mathbf{W}_{q,p}$, $\mathbf{W}_a$ is referred to as an alternating function. In the following, we first provide definitions for the gradient (grad) and curl operators, and then introduce the Hodge Decomposition Theorem.

Definition 1 (Gradient Operator). *The 'grad' operator measures the flow of influence between nodes. For any edge $e = (p, q) \in E$, it is defined as:*

$$grad(s) = s_q - s_p, \tag{4}$$

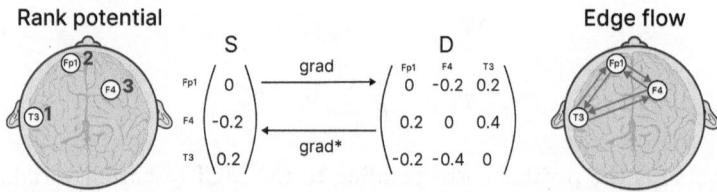

Fig. 2. Node ranking in a graph involves a gradient operator that maps ranking scores on the left to gradient flow on the right (dependence structures). The adjoint, gradient star, reverses this mapping from right to left.

where s_p and s_q represent the scores or potentials of nodes p and q. This operator reflects the change in influence from node p to node q.

Definition 2 (Curl Operator). *The 'curl' operator measures the rotation strength of mutual influence among nodes. For any face $t = (p,q,r) \in T$, it is defined as:*

$$curl(\phi) = \phi_{p,q} + \phi_{q,r} + \phi_{r,p}, \tag{5}$$

where $\phi_{p,q}$, $\phi_{q,r}$, and $\phi_{r,p}$ are the edge flows between nodes p, q, and r. This operator evaluates the total rotation around a triangular face in the graph.

Given the definitions above, we can observe that the image of the gradient operator is included in the kernel of the curl operator. This relationship leads to the Helmholtz-Hodge Decomposition theorem [3].

Theorem 1 (Helmholtz-Hodge Decomposition). *The space of anti-symmetric matrices, $\mathbb{R}^E$, can be decomposed into three mutually orthogonal subspaces:*

$$\mathbb{R}^E = Im(grad) \oplus Im(curl^*) \oplus Ker(grad^*) \cap Ker(curl).$$

The anti-symmetric component $\mathbf{W}_a$ can exhibit cyclic influence in the network, which indicates that no perfect ranking can be achieved without violating at least one mutual influence relationship within each cycle. According to the Helmholtz-Hodge Decomposition theorem, any edge flow $\mathbf{W}_a$ can be uniquely decomposed as:

$$\mathbf{W}_a = -grad(s) + curl^*(\phi) + \mathbf{h}, \tag{6}$$

where $grad(s)$ is the gradient component capturing hierarchical relationships, $curl^*(\phi)$ is the curl component indicating local cyclic influences, and $\mathbf{h}$ is the harmonic component indicating non-local or global cyclic influences.

To estimate these components, we use the least squares method for the gradient and curl components, and obtain the harmonic component as the residual.

$$\hat{s} = \arg\min_s \|\mathbf{W}_a + grad(s)\|_F^2, \tag{7}$$

$$\hat{\phi} = \arg\min_\phi \|\mathbf{W}_a - curl^*(\phi)\|_F^2, \tag{8}$$

$$\hat{h} = \mathbf{W}_a + grad(s) - curl^*(\phi). \tag{9}$$

The estimated score vector $\hat{s} = [s_1, s_2, \ldots, s_N]'$ (where N is the number of nodes or time series components) derived from $\mathbf{W}_a$ is used to establish the topological ranking of the EEG channels (or brain regions). By minimizing the difference between $-\mathrm{grad}(\hat{s})$ and $\mathbf{W}_a$, this ranking accurately reflects net information flow patterns while avoiding inconsistencies, as $\mathrm{curl}(\mathrm{grad}(\hat{s})) = 0$. If $s_p > s_q$, the p-th time series tends to lead other time series more than the q-th time series, indicating that p is higher in the hierarchy. Thus, $\hat{s}$ captures the brain hierarchy, as illustrated in Fig. 2. This is crucial for understanding brain function and the impact of dysfunctions such as seizures.

Our methodology involves a series of steps designed to rank brain regions based on their mutual influence. Initially, we employ the PCMCI method to estimate effective brain connectivity, utilizing causal inference techniques. Following this, the connectivity network is decomposed into symmetric and anti-symmetric components. We then apply Hodge Decomposition to the anti-symmetric component, extracting the gradient, which is essential for establishing the ranking. This allows us to derive the topological ranking of brain regions based on the gradient component. The entire process is visually represented in Fig. 3.

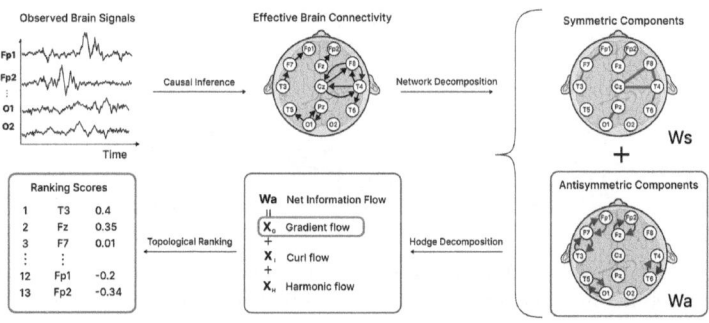

Fig. 3. Hodge decomposition pipeline for topological ranking of brain regions.

The computational complexity is primarily driven by the PCMCI approach, which includes a two-stage process. The first stage, using the PC algorithm, can have complexity up to $O(P^d)$ in the best case and potentially exponential in the worst case, where d is the maximum number of parent nodes. The second stage, the Momentary Conditional Independence (MCI) test, has complexity $O(L \times P^2)$, with L as the number of time lags and P the number of channels. The subsequent decomposition into symmetric and anti-symmetric components requires $O(P^2)$, and the Hodge Decomposition involves solving a least squares problem, that is equivalent to solving a linear system of size P in $O(P^3)$. These latter steps are efficient for low-dimensional data, such as EEG datasets, with PCMCI being the main computational bottleneck. For high-dimensional data, alternative methods may be needed to ensure efficient and accurate connectivity estimation.

3   Experiments

To evaluate our methodology, we propose six distinct simulation scenarios, each representing different hierarchical dependency structures. These scenarios range from linear hierarchies to complex cyclic dependencies, testing the robustness and versatility of our ranking mechanism across various topologies, as illustrated in Fig. 4.

- **Scenarios 1 and 4:** Exhibit a linear hierarchical dependence structure, allowing for clear node ranking (scenario 4 reverses scenario 1).
- **Scenarios 2 and 5:** Feature cyclic dependence structures, challenging meaningful ranking due to non-transitivity (scenario 5 reverses scenario 2).
- **Scenarios 3 and 6:** Present partial rankings with localized cycles, suggesting intermediate dependence structures (scenario 6 reverses scenario 3).

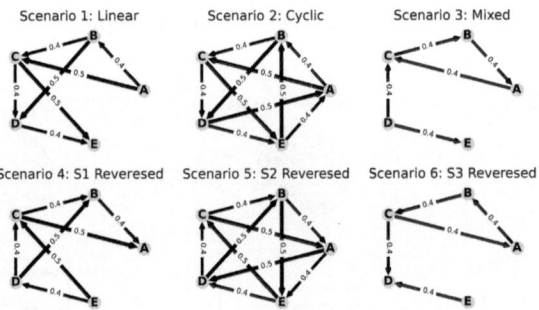

Fig. 4. Visual representation of the designed dependency networks for six simulation scenarios. Each arrow indicates a dependency direction at lag one.

We generate multivariate time series data using the mixture model in Eq. 10, with $P = 5$ dimensions and $T = 30,000$ observations, divided into 500-observation epochs to analyze temporal stability and variability in ranking. The first 5,000 observations correspond to scenario one, the next 5,000 to scenario two, and so on.

$$Y(t) = \Phi(t)Z(t) + E(t), \quad \Phi(t) = \begin{bmatrix} 0 & 0.4 & 0.5 & 0 & 0 \\ 0 & 0 & 0.4 & 0.5 & 0 \\ 0 & 0 & 0 & 0.4 & 0.5 \\ 0 & 0 & 0 & 0 & 0.4 \\ 0 & 0 & 0 & 0 & 0 \end{bmatrix}, \tag{10}$$

where $Y(t)$ is the observed multivariate time series, $Z(t)$ is a multivariate standard Gaussian process, and $E(t)$ is random noise, also modeled as a standard Gaussian process. The mixing matrix $\Phi(t)$ reflects each scenario's dependence structure.

Our method follows an epoch-based approach to dynamically estimate causal relationships within the multivariate time series data in order to capture the directional dependencies and hierarchical variations. Figure 5 showcases the causal connections identified for one epoch from scenario one, highlighting both the causal pathways and the strength of these interactions.

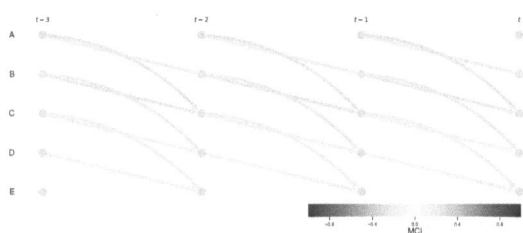

Fig. 5. Visual representation of the causal interactions within the network for scenario one as identified by PCMCI, showcasing the algorithm's capacity to accurately capture both the direction and magnitude of causal dependencies across time lags.

By applying our proposed ranking procedure, we are able to identify and highlight nodes within the network that exert significant influence over others, providing a clear understanding of the hierarchical patterns and their dynamics, as can be seen in Fig. 6. Our methodology effectively ranks the simulated time series, consistently placing node A at the top and node E at the bottom, as indicated by their respective curves, in scenario one, with the inverse pattern observed in scenario four. A node is ranked above another if its curve is consistently above the other's. Scenarios two and five exhibit fluctuating rankings due to inherent cyclic dependencies, while scenarios three and six show partial rankings with notable positions for node D, aligning with the designed simulation scenarios. These results validate our approach for ranking multivariate time series data based on net influence, as it robustly identifies the hierarchical structures in scenarios one and three. In contrast, scenarios two and four, characterized by cyclic dependencies, exhibit inconsistent rankings. Through these

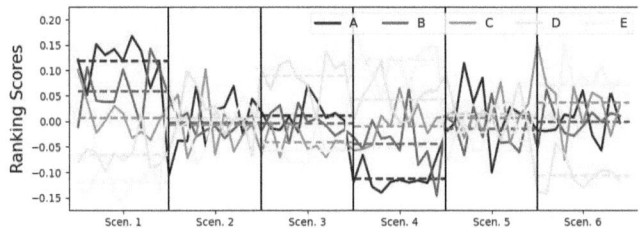

Fig. 6. Aggregate causality-based ranking scores, delineating the hierarchical structure within each scenario. The distinct patterns observed across scenarios validate the robustness of our ranking methodology.

simulations, we confirm that our CBTR method effectively uncovers the hierarchical structure within dependence networks, distinguishing various dependency patterns.

4   Results

In this section, we focus on applying our methodology to a dataset of neonatal EEG recordings [24]. Neonatal seizures are a critical concern within the neonatal intensive care unit (NICU), raising numerous questions about their temporal and spatial characteristics. The dataset comprises 19-channel EEG recordings from 79 full-term infants at the Helsinki University Hospital NICU, with seizure annotations provided by three expert clinicians for each recording, spanning a median duration of 74 min. We concentrate our study on a few subjects, specifically analyzing epochs consistently classified as either seizure or seizure-free by all three experts. Our initial goal is to explore how seizures influence effective brain connectivity in neonates, serving as a foundational step for future detailed research on the nature of seizures and their developmental effects.

In many instances, neonatal EEG recordings reveal distinct seizure patterns, as shown in Fig. 7. However, deciphering the interconnections among the 19 channels, particularly the influence of seizures, presents a complex challenge. Our approach quantifies the impact of seizures on the neonatal brain's effective connectivity at the channel level, aiming to uncover how seizures modify brain connectivity and hierarchical organization in each subject. The EEG recordings typically exhibit smooth variations at the millisecond scale, leading to high autocorrelation in lagged observations, which can increase spurious causal relationships. To mitigate this, we analyze effective brain connectivity through successive differences or increments in EEG recordings, reducing the influence of lagged dependencies, as illustrated in Fig. 7.

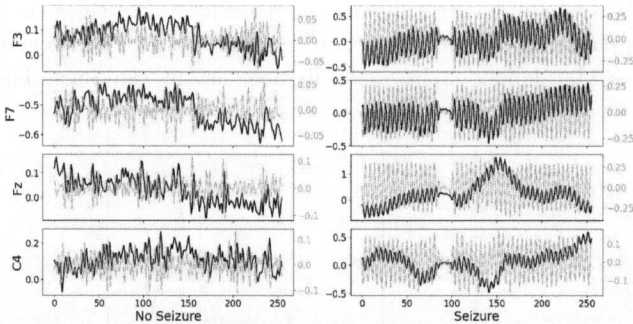

Fig. 7. Sample of EEG signals from channels F3, F7, Fz, and C4, showcasing epochs (1 s each) with and without seizure. Original EEG recordings are depicted in black, while the incremental changes between successive EEG readings are highlighted in cyan. (Color figure online)

Following the approach presented in Sect. 2.1, we estimated the causal influences among brain channels for each subject across different time lags during seizure and non-seizure periods. Following the procedure in Sect. 2.2, we evaluate the changes in causal influence during seizure time by estimating ranking scores for all channels across 50 seizure epochs and compare them with scores from 50 seizure-free epochs. We used a t-test to determine the statistical significance of the ranking score differences. This analysis revealed significant changes in causality-based ranking scores for several channels for each examined subject, evidenced by very small p-values. For instance, in subjects 1 and 5, we highlight the EEG channels most significantly impacted by seizures. These findings, depicted in Figs. 8 and 9, illuminate how seizures alter the hierarchical architecture of brain connectivity, offering deeper insights into their impacts and pinpointing the brain regions where seizures exert the greatest influence. These results demonstrate the efficacy of our CBTR approach in uncovering the nuanced dynamics of effective connectivity influenced by seizure events.

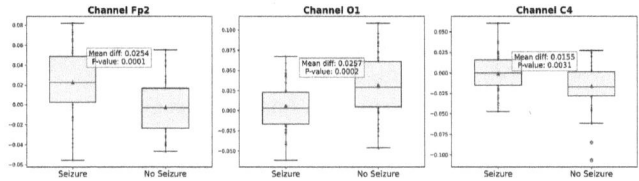

Fig. 8. Channels with most significant shifts in the ranking scores between seizure free epochs and seizure epochs. For subject 1.

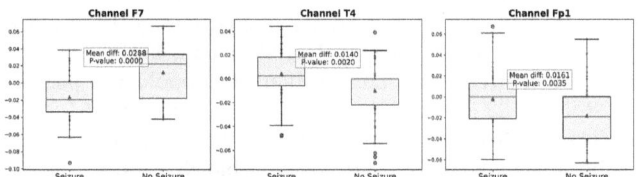

Fig. 9. Channels with most significant shifts in the ranking scores between seizure free epochs and seizure epochs. For subject 5.

5   Conclusion and Discussion

In this paper, we introduced a novel framework for investigating effective brain connectivity, integrating causal inference methods with Hodge decomposition to rank brain regions based on their mutual causal influences. This novel approach unlike traditional methods that focus on functional connectivity via persistence homology, analyzes the oriented dependencies of effective brain connectivity.

Our simulation study showcased the robustness and accuracy of our methodology in uncovering and describing hierarchical structures within multivariate dependence networks. Additionally, applying our method to neonatal EEG data revealed valuable insights into how seizures impact brain connectivity, particularly highlighting changes in the hierarchical positions of specific brain regions. Future work will delve deeper into the analysis of epileptic seizures, leveraging our method's potential to improve seizure localization.

References

1. Adams, H., Emerson, T., Kirby, M., Neville, R., Peterson, C., Shipman, P.: Persistence images: A stable vector representation of persistent homology. Journal of Machine Learning Research **18**, 1–35 (2017). https://jmlr.org/papers/volume18/16-337/16-337.pdf
2. Bendich, P., Marron, J.S., Miller, E., Pieloch, A., Skwerer, S.: Persistent homology analysis of brain artery trees. The Annals of Applied Statistics **10**, 198–218 (2016). https://doi.org/10.1214/15-AOAS886
3. Bhatia, H., Norgard, G., Pascucci, V., Bremer, P.T.: The helmholtz-hodge decomposition-a survey. IEEE Trans. Visual Comput. Graphics **19**(8), 1386–1404 (2013). https://doi.org/10.1109/TVCG.2012.316
4. Bromfield, E.B., Cavazos, J.E., Sirven, J.I.: An Introduction to Epilepsy. American Epilepsy Society (2006)
5. Bubenik, P.: Statistical topological data analysis using persistence landscapes. J. Mach. Learn. Res. **16**, 77–102 (2015). https://doi.org/10.5555/2789272.2789275
6. Carlsson, G., Zomorodian, A., Collins, A., Guibas, L.: Persistence barcodes for shapes. p. 124-135. Association for Computing Machinery (2004). https://doi.org/10.1145/1057432.1057449
7. Cook, C.J., et al.: Effective connectivity within the default mode network in left temporal lobe epilepsy: findings from the epilepsy connectome project. Brain Connectivity **9** (2019). https://doi.org/10.1089/brain.2018.0600
8. Edelsbrunner, H., Harer, J.: Persistent homology-a survey. Discret. Comput. Geom. **453**, 257–282 (2008). https://doi.org/10.1090/conm/453/08802
9. Edelsbrunner, H., Letscher, D., Zomorodian, A.: Topological persistence and simplification **28**, 511–533 (2002). https://doi.org/10.1007/s00454-002-2885-2
10. El-Yaagoubi, A.B., Jiao, S., Chung, M.K., Ombao, H.: Spectral topological data analysis of brain signals. arXiv:2401.05343 [q-bio.NC] (2024). https://doi.org/10.48550/arXiv.2401.05343
11. El-Yaagoubi, A.B., Ombao, H.: Topological data analysis for directed dependence networks of multivariate time series data. In: Research Papers in Statistical Inference for Time Series and Related Models, chap. 17. Springer (2023)
12. Friston, K., Moran, R., Seth, A.: Analysing connectivity with granger causality and dynamic causal modelling. Curr. Opin. Neurobiol. **23**(2), 172–178 (2013). https://doi.org/10.1016/j.conb.2012.11.010
13. Friston, K.J.: Functional and effective connectivity: A review. Brain Connectivity **1**, 13–36 (2011). https://doi.org/10.1089/brain.2011.0008
14. Gidea, M., Katz, Y.: Topological data analysis of financial time series: Landscapes of crashes. Phys. A **491**, 820–834 (2018). https://doi.org/10.1016/j.physa.2017.09.028
15. Jiang, X., Lim, L.H., Yao, Y., Ye, Y.: Statistical ranking and combinatorial hodge theory. Math. Program. **127**(1), 203–244 (2011). https://doi.org/10.1007/s10107-010-0419-x

16. Jiang, Y., Chen, D., Chen, X., Li, T., Wei, G.W., Pan, F.: Topological representations of crystalline compounds for the machine-learning prediction of materials properties. npj Computational Materials **7**(28) (2021). https://doi.org/10.1038/s41524-021-00493-w

17. Lee, H., Kang, H., Chung, M.K., Kim, B.N., Lee, D.S.: Persistent brain network homology from the perspective of dendrogram. IEEE Trans. Med. Imaging **31**, 2267–2277 (2012). https://doi.org/10.1109/TMI.2012.2219590

18. Pearl, J.: Causality: Models. Cambridge University Press, Reasoning and Inference (2009)

19. Rabadan, R., Blumberg, A.J.: Topological Data Analysis for Genomics and Evolution. Cambridge University Press (2019)

20. Roebroeck, A., Formisano, E., Goebel, R.: Mapping directed influence over the brain using granger causality and fmri. Neuroimage **25**(1), 230–242 (2005). https://doi.org/10.1016/j.neuroimage.2004.11.017

21. Runge, J.: Discovering contemporaneous and lagged causal relations in autocorrelated nonlinear time series datasets. In: Peters, J., Sontag, D. (eds.) Proceedings of the 36th Conference on Uncertainty in Artificial Intelligence (UAI). Proceedings of Machine Learning Research, vol. 124, pp. 1388–1397. PMLR (03–06 Aug 2020), https://proceedings.mlr.press/v124/runge20a.html

22. Runge, J., Nowack, P., Kretschmer, M., Flaxman, S., Sejdinovic, D.: Detecting and quantifying causal associations in large nonlinear time series datasets. Science Advances **5**(11), eaau4996 (2019). https://doi.org/10.1126/sciadv.aau4996

23. Seth, A.K., Barrett, A.B., Barnett, L.: Granger causality analysis in neuroscience and neuroimaging. J. Neurosci. **35**(8), 3293–3297 (2015). https://doi.org/10.1523/JNEUROSCI.4399-14.2015

24. Stevenson, N.J., Tapani, K., Lauronen, L., Vanhatalo, S.: A dataset of neonatal eeg recordings with seizure annotations. Scientific Data **6**, 190039 (2019). https://doi.org/10.1038/sdata.2019.39

25. Weichwald, S., Peters, J.: Causality in cognitive neuroscience: Concepts, challenges, and distributional robustness. J. Cogn. Neurosci. **33**(2), 226–247 (2021). https://doi.org/10.1162/jocn_a_01623

Author Index